Low-Temperature Technologies and Applications

Edited by Salim Newaz Kazi

Published in London, United Kingdom

IntechOpen

Supporting open minds since 2005

Low-Temperature Technologies and Applications
http://dx.doi.org/10.5772/intechopen.87698
Edited by Salim Newaz Kazi

Contributors
Jeom Kee Paik, Dong Hun Lee, Jonas W. W Ringsberg, P. J. Tan, Gérard Ghibaudo, Mikaël Cassé, S.B. Vishnu, Biju T. Kuzhiveli, Solomon O. Banjo, Bukola O. Bolaji, Oluseyi O. Ajayi, Olatunde A. Ajani Oyelaran, Selva Pandian Ebenezer, Salim Newaz Kazi

First published in London, United Kingdom, 2022 by IntechOpen
IntechOpen is the global imprint of INTECHOPEN LIMITED, registered in England and Wales, registration number: 11086078, 5 Princes Gate Court, London, SW7 2QJ, United Kingdom
Printed in Croatia

British Library Cataloguing-in-Publication Data
A catalogue record for this book is available from the British Library

Additional hard and PDF copies can be obtained from orders@intechopen.com

Low-Temperature Technologies and Applications
Edited by Salim Newaz Kazi
p. cm.
Print ISBN 978-1-83968-584-2
Online ISBN 978-1-83968-585-9
eBook (PDF) ISBN 978-1-83968-586-6

For EU product safety concerns:
IN TECH d.o.o., Prolaz Marije Krucifikse Kozulić 3, 51000 Rijeka, Croatia, info@intechopen.com or visit our website at intechopen.com.

Meet the editor

Dr. S. N. Kazi is a Professor of Mechanical Engineering at the University of Malaya. He has a specialization in Heat Transfer, Fluid Mechanics, Particle Characterization, Heat Exchanger Fouling Mitigation, Nanofluid synthesis and applications, and Renewable Energy. He has a long Engineering service experience in Petrochemical Industries. He also worked as a consultant for different Engineering Companies. Dr. Kazi has an academic background with the B. Sc., M. Sc., and M. S. in Mechanical Engineering. He received his Ph.D. in Chemical and Materials Engineering. He is a member of many professional bodies and an editorial member of many journals. He has been working as an academic since 2009. He has completed supervision of many postgraduate theses, published many technical papers, and edited some books.

Contents

Preface **XI**

Section 1
Low-Temperature Technologies and Modelling **1**

Chapter 1 **3**
Introductory Chapter: Low-Temperature Technologies
by Salim Newaz Kazi

Chapter 2 **17**
Low Temperature Characterization and Modeling of FDSOI Transistors for Cryo CMOS Applications
by Mikaël Cassé and Gérard Ghibaudo

Chapter 3 **47**
Effect of Roughness Elements on the Evolution of Thermal Stratification in a Cryogenic Propellant Tank
by S.B. Vishnu and Biju T. Kuzhiveli

Chapter 4 **67**
Ultimate Compressive Strength of Steel Stiffened-Plate Structures Triggered by Brittle Fracture under Cryogenic Conditions
by Dong Hun Lee, Jeom Kee Paik, Jonas W. Ringsberg and P.J. Tan

Section 2
Low-Temperature Refrigeration and Applications **91**

Chapter 5 **93**
Refrigerant Mixtures
by Selva Pandian Ebenezer

Chapter 6 **111**
Impact of Working Fluids and Performance of Isobutane in the Refrigeration System
by Solomon O. Banjo, Bukola O. Bolaji, Oluseyi O. Ajayi and Olatunde A. Oyelaran

Preface

This book discusses aspects, applications, resources, and deployment of low-temperature technology. It describes azeotropic, non-azeotropic/zeotropic, and their mixtures in refrigeration for gliding evaporation and condensation along with applications are elaborately discussed in this book. It also discusses the modeling, design, and augmentation of low-temperature cooling processes.

Generation of plasmas (LTP) and its application in low temperatures are highlighted in this book. Discussions are elaborated on the efficient production of the radicals by the energetic electrons. Excited states, photons, and charged species are also incorporated in the book. The mechanism of accelerated space-charge sheaths which accelerate and deliver fluxes of ions to surfaces with adjustable energies are discussed. Discussion on material surface alteration by ion fluxes by sputtering, activation, etching, and deposition that are crucial to technological devices are incorporated. The application of plasma in different research fields in science and engineering is elaborately discussed in this book as well as the application of plasma from different generating devices. Detailed discussion on societal benefits of low-temperature plasmas and their wide range of applications are a part of this book as well in this book.

Preservation of metallic structures exposed suddenly to low temperatures is discussed in this book. Application of steel stiffened plate structures, along with ultimate stress analysis, brittle fracture of stiffened plates, rupture, ductile fracture of materials is incorporated in the analysis. In this book, a detailed discussion has been incorporated for the understanding and application of the structural alterations to sustain in a cryogenic environment.

Vapor compression refrigeration system (VCRS) is also a theme in this book, along with the consideration of thermo-physical and thermodynamics properties, ecofriendly nature, and economic aspects of selection of refrigerants.

I am grateful to the editorial team members for their support. I am also thankful to my wife Nilufa Parveen and my son Kazi Mehrab Newaz for supporting me throughout the preparation of this book.

Salim Newaz Kazi
Faculty of Engineering,
Department of Mechanical Engineering,
University of Malaya,
Kuala Lumpur, Malaysia

Section 1

Low-Temperature Technologies and Modelling

Chapter 1

Introductory Chapter: Low-Temperature Technologies

Salim Newaz Kazi

1. Introduction

This book has introduced low-temperature technology, its aspects, and applications. Further, it has covered low-temperature resources and their deployment technologies. Sterilization and preservation techniques and their engineering and scientific aspects are incorporated. Ultra-low-temperature refrigeration is a crucial requirement in some research and industrial applications, which is considered in depth in this issue. Azeotropic, zeotropic, and non-zeotropic refrigerants and their mixtures are highlighted in detail along with their applications and economic aspects. Cryogenics, low temperature, and vacuum systems for industrial applications are incorporated. Global energy resources, energy crisis, the principle of conservation of energy, alternative sources of energy for low-temperature technologies, and their application in refrigeration, process industries, and electronic equipment and parts manufacturing industries are considered here. Enhancement of efficiency of the system along with the economic aspects is incorporated. At the end, modeling, design efficiency, and enhancement of efficiency of low-temperature cooling process are elaborately discussed.

2. Overview of low-temperature technologies

In a refrigeration system, the used refrigerant could be either a pure compound or a mixture (blend) of two or more compounds or refrigerants. R12, R22, and R134a are the pure refrigerants, whereas the mixture refrigerants are R502, R404A, and R407C [1]. A mixture of compounds used in a refrigeration loop can behave either as a pure refrigerant (azeotropic mixtures), or differently (non-azeotropic or zeotropic, mixtures). Azeotropic mixtures contain two or more refrigerants and the mixture at a certain pressure evaporates and condenses at a constant temperature. It behaves like pure refrigerants in all practical aspects. For a given pressure, the temperature is constant in the liquid-vapor mixture region. Among the non-azeotropic/zeotropic mixtures, the zeotropic mixtures have gliding evaporation and condensing temperature. At the evaporation temperature, the most volatile component boils off first and the least volatile component boils off last. The opposite happens when a gas condenses into liquids and the temperatures changes in the liquid-vapor mixture at a given pressure. It caused a raised evaporation and condensing temperature along the heat transfer surface. In general, there is a negative effect on the heat transfer coefficient due to mass transport phenomena in the refrigerant during the evaporation and condensation of zeotropic. This negative effect is neutralized by the high turbulence and good mixing in the heat exchanger, which also suggests that Brazed Plate Heat Exchangers (BPHEs) have an advantage over Shell and Tube Heat Exchangers (S&Ts). If the refrigerant liquid is allowed to accumulate somewhere in the circuit, then the problems arise at the suction line of the accumulators, flash tanks, receivers, or pool boiling/flooded evaporators (often S&T) [1]. The unpredictable performance could be observed from a change in the composition of the refrigerant circulating through the system. Thus, all components should have a continuous flow of refrigerant to avoid unpredictable performance. Thus, for systems containing mixtures with glide, the flooded evaporators will probably disappear. A considerable leak of refrigerant in the liquid-vapor region can lead to the same problems, as appears in the case of the changed composition of the remaining refrigerant charge, which produces unpredictable system performance.

Plasmas (low-temperature plasmas [LTPs]) at low temperature have characteristic electron energies of a few eV–10 eV with ionization of typically lower degrees but they can reach tens of percent in arc discharges. These energetic electrons can efficiently generate radicals, excited states, photons, and charged species. Particularly at low pressure, the space charge sheaths at the boundary of plasmas accelerate and deliver fluxes of ions to surfaces with adjustable energies ranging from a few to hundreds of eVs. Surface modifications are enabled by these ion fluxes, which alter surfaces by sputtering, activation, etching, and deposition that are essential to technological devices. Many industrial applications are performed by plasma created by microwave, arc, and inductively coupled plasma discharges that operate close to thermal equilibrium [2, 3].

Most of the low-temperature plasmas substantially deviate from thermodynamic equilibrium, along with the electron temperature T_e being significantly higher than the heavy particle temperature and gas temperature T_g. A chemically rich environment could be generated by LTP sources at close to room temperature both at reduced and at ambient pressures, which is an exclusive condition that facilitates the delivery of highly reactive plasma species in a non-destructive and useful way to extremely heat-sensitive surfaces. The microelectronics industry formed the technological base of modern society. Currently, the beneficial plasma-surface interactions deposited and removed the nanometer resolution in the fabrication of microprocessors [4]. This useful contact with surfaces later extended to liquids, organic tissues, and wounds, which directed to the arena of plasma medicine [5]. There could be an interaction between LTPs and surfaces with the plasma, non-destructively and beneficially [6]. They synthesized nanomaterials as particles or aerosol-laden dusty plasma that assisted nanomaterial synthesis. It was observed that LTPs can be generated and sustained within liquids and bubbles in liquids. Currently, they are investigated for medical applications, chemical processing, and in the context of environmental stewardship [7]. Thus, there are enormous extraordinary societal benefits of low-temperature plasmas. Low-temperature plasmas are remarkably interdisciplinary and have a wide range of applications. LTP discipline joined varieties of research fields, such as fluid dynamics, electrodynamics, thermodynamics, heat transfer, statistical physics, atomic and molecular physics, material and surface science, chemistry, electrical engineering, chemical engineering, and recently even biology and medicine.

Considering the enormous diversity in applications, investigations on non-equilibrium plasma kinetics and the interactions of plasmas with the matter should be encouraged. Research on Plasma Agriculture and Innovative, Plasmas in Analytical Chemistry, Plasma Metamaterials, and Plasma Photonic Crystals are the new additions in this activity area. Applications of plasmas in the areas of energy, flow control and material processing, synthesis, transport, and catalysis are also highlighted in many reviews. There are scopes of study on the dominant mode of energy transfer and chemical reaction processes in transient plasmas, the physical and chemical interaction of plasmas with materials and liquids, and the mechanisms, origins of the formation of complex self-organizing structures in plasmas, etc. Research on the low-temperature plasma field is at stake. It is observed that the support of fundamental research in the past has provided good results. Application of low-temperature plasmas depends on culturing and maintaining new generations of researchers engaged in plasma science, modeling, and diagnostics. Thus, more involvement should be encouraged in the fundamentals of plasma science that strengthens this technological development and enables the career progress of the next generation [8].

Application of steel-stiffened panels is commonly used in offshore, navel, mechanical, and civil engineering structures. Sometimes, they get in contact with cryogenic conditions, such as the sudden release of liquefied gas, such as LNG (liquefied natural gas) or liquefied hydrogen [9]. For the design of steel-stiffened plate structures, the ultimate strength is a primary criterion. Thus, it is necessary to characterize the effects at cryogenic conditions on the ultimate strength of such structures [10, 11]. Some works introduced a sequel to examine under cryogenic conditions, the brittle fracture of stiffened-plate steel structures. On some occasions, rupture, ductile fracture, and brittle fracture were investigated. In the case of fracture, the inherently brittle materials show different behavior than the ductile materials such as carbon steel. Slow and stable crack growth is generally exhibited during crack extension in the ductile materials. Similar behavior to brittle materials could be seen in certain environments such as very low temperatures or lower than the ductile-to-brittle fracture transition temperatures (DBTT), impact loading etc. Here, the Bauschinger effect of materials cannot be neglected at cryogenic conditions because the material behavior in compression is distinct from that in tension [10, 12, 13]. Some constitutive equations of materials were proposed to compute the failure behavior of structures at cold (sub-zero) temperatures (or higher than the ductile-to-brittle fracture transition temperatures). It is identified that there is a lack of investigation applicable to entirely brittle fracture at cryogenic conditions [14–17]. The research compared the ultimate compressive strength of steel-stiffened plate structures by nonlinear finite element method (NLFEM) using the multi-physics software package LS-DYNA implicit code. The tested structure was fabricated from high-strength steel of grade AH32. Mechanical properties were considered, tension and compression tests at low temperatures and cryogenic conditions, and later, a phenomenological relation of engineering stress versus engineering strain of the material was formulated, and then, the model was used into the LS-DYNA implicit code. Comparison results of the NLFEM model data with the experimental results from the full-scale test presented the similarity of the trends [18, 19].

At temperatures higher than the temperature of the ductile-to-brittle fracture transition, the ductile-dominated behavior of structural steel is observed. The material behaves predominantly in a brittle manner with partial or no ductility at the approaching cryogenic condition. Ductile-to-brittle fracture transition temperature (DBTT) has to be clarified further. However, it is reported in a few investigations the evidence for brittle fracture behavior of steel structures at the cryogenic condition and their dependence on the type of materials and loading conditions (e.g., quasi-static or impact), in addition to other relevant factors. In some investigations, the ductile fracture in steel tubes under quasi-static loads at −60°C during the crashing test was observed. The brittle fracture was reported in dropped object impact testing of steel-stiffened plate panels at −60°C. Collapse testing of a steel-stiffened plate structure under axial-compressive loading in the presence of brittle fracture the ultimate strength was reached. In the presence of flexural-torsional buckling at room temperature, the structures reached the ultimate strength. On the other hand, brittle fracture caused the global failure of the material structure at cryogenic conditions.

In this book, it has attempted to develop a new fracture criterion based on the hypothesis that crack initiates if an equivalent stress exceeds a critical value. In this book, the effects of brittle fracture on the ultimate compressive strength of stiffened plate steel structures under cryogenic conditions have thoroughly been investigated. To analyze the ultimate compressive strength of steel-stiffened plate structures, the developed computational models were applied considering brittle fracture, under cryogenic conditions. Here, a material model for the high-strength steel at cryogenic condition is proposed, considering the Bauschinger effect, and implemented into a nonlinear finite element solver (LS-DYNA). The researchers obtained a good agreement between computational predictions and experimental measurements for the ultimate compressive strength response of a full-scale stiffened plate steel structure. Thus, a practical method was introduced to compute the ultimate compressive strength of steel-stiffened plate structures at cryogenic conditions triggered by brittle fracture. A recommendation suited for the adoption of NLFEM (nonlinear finite element method) simulations.

In the refrigeration process, the low-temperature technology is achievable, especially in the vapor compression refrigeration system (VCRS) [20, 21]. With reference to the issue of environmental challenges, the appropriate selection of refrigerant is essential to protect the environment from the greenhouse effect and stratospheric ozone layer depletion. Based on their safety, thermo-physical and thermodynamics properties, and economic factors, the refrigerants can be selected [22, 23].

As a refrigerant, ammonia has an increased coefficient of performance at a low cost, favorable transport, high enthalpy of vaporization, and good thermodynamic properties. Due to toxicity, it is prevented for domestic purposes [24–26]. Later, halocarbon refrigerants were developed. They have high thermal efficiency but imposed a threat to the ozone layer. Finally, the hydro-chlorofluorocarbon refrigerants were found as the best short-term alternative replacement for chlorofluorocarbon. Later, its application was also becoming limited. Du Pont said the HCFC substances' production should be ceased but would persist in making it available for existing equipment until their expiry period [27–29].

The hydrofluorocarbon refrigerant (R134a) has zero ODP with a high GWP of 1430, which promotes its application as a working fluid that creates a threat to the immediate surroundings. The challenge of climate change engaged the researchers toward discovering and applications of environmentally friendly refrigerants in the heating and air-conditioning systems that also diminish greenhouse gas and enable energy savings [30–32].

Regulatory bodies (Kyoto and Montreal protocols) have called for banning pure fluid, which is posing a threat of high global warming and ozone depletion [33–35]. Halogen-free refrigerants (HFRs) are an alternative to halocarbon refrigerants in the refrigeration system. They are eco-friendly refrigerants with an organic composition of hydrogen and carbon atoms, such as butane, propane, isobutene, propylene [23]. It is miscible with mineral oil, which provides a smooth running of the single hermetic reciprocating compressor (SHRC). It has got zero ozone depletion and negligible global warming potential. The critical temperature of HFR refrigerant is high that enhances the domestic refrigerator's efficiency. It preserves the environment and serves as an energy reduction substance [36–42].

In a closed system, the halogen-free refrigerant is applicable with relatively less mass charge. It is used in significant engineering systems such as deep freezers, water dispensers, domestic refrigerators, vending machines, and industrial refrigeration system. Some investigators used isobutane (R600a) and hydrofluorocarbon (R134a) as refrigerants and compared their performance characteristics. It was observed that the role of HTM (working fluid) in the heating ventilation and air-conditioning industries are highly significant [23, 43].

Considering the thermodynamic properties, the conventional refrigerants have predominantly been used. Hydrocarbon refrigerants such as isobutane were later picked because of their negligible global warming potential, zero ozone depletion, and eco-friendly behavior. It was reported that a better absorption of refrigerant into a refrigeration system could increase the coefficient of performance (COP). By using R600a and R134a, the performance was raised. Researchers have developed acceptable suggestions for working fluid (refrigerant) that would be suitable for a domestic refrigeration system. It could also enable the cooling performance of a computation system at a lower minimum level. Based on the comparative performance of R600a and R134a refrigerants in the vapor compression refrigeration system, the overall efficiency and refrigeration effect of the cooling system are greatly influenced by the working fluids. The researchers reiterated that the use of isobutane could secure the surroundings from global warming, ozone depletion, improvement of energy conservation etc.

The application of cryogenic fluids for keeping temperature and maintaining electronic circuits was successfully demonstrated by many authors with the emerging field of quantum computing, and the study of CMOS devices at low and very low temperatures, below 100 K, has received a revived attention [44–48]. Recently, in advanced CMOS technologies, outstanding characteristics were demonstrated at 4.2 K [49–52], specifically for Fully Depleted Silicon-On-Insulator (FDSOI) [53–55]. Low-power electronics, threshold voltage tune ability, and back bias ability were designed by applying FDSOI, which offered low variability due to the un-doped channel [56]. The low-temperature operation can provide many advantages such as better electrical performance of MOSFETs, higher carrier drift velocity, higher on-state drain current and transconductance, steeper subthreshold slope, lower leakage current, etc. [45, 52]. The temperature dependence of carrier transport properties and thermal effects needed to be incorporated to obtain accurate models [57, 58]. Due to low-temperature operation, the appeared physical phenomena need to be properly modeled [52]. This book presents a review of recent results obtained on 28-nm FDSOI transistors operated down to deep cryogenic temperatures. The major device electrical properties in terms of transfer characteristics and MOSFET parameters as a function of temperature are elaborately discussed. It is reported that the self-heating phenomena could alter the performance of the FDSOI device. It also contains the matching and variability properties of scaled transistors and their limitations in the analog applications. It presented the development of a compact model necessary for FDSOI circuit design at deep cryogenic temperatures and the operation of elementary circuits at very low temperatures regarding inverter delay and oscillator frequency.

The investigation of CMOS technology performance at cryogenic temperatures was encouraged for its applications in high-performance computing and energy physics. The advanced CMOS node could support the larger bandwidth at low-temperature applications of quantum computers. For co-integration between qubits and consistent engineering of control and read-out, the FDSOI technology appears as a valuable solution. For the advanced CMOS node behavior at deep cryogenic operation, more investigations are required. Thus, this book contains a review of electrical characterization and modeling results recently obtained on modern FDSOI MOSFETs down to 4.2 K.

Detail of the self-healing phenomena was characterized and provided valuable information about the actual device temperature as a function of power dissipation and the thermal resistance that limits the heat dissipation in the FDSOI structure specifically at low temperatures. Considering the threshold voltage and drain current statistical variability analysis, the matching properties were studied and enlightening the mismatch in FDSOI transistors that increases about 30–40% at deep-cryogenic temperatures. On the other hand, Poisson-Schrodinger simulations were carried out with success down to zero Kelvin, giving access to valuable information about the gate charge control in FDSOI structures as a function of temperature, which provided physical insight to the development of compact model mandatory for FDSOI circuit design at deep cryogenic temperatures. The operation of elementary circuits such as ring oscillators and voltage-controlled oscillators was validated in terms of inverter delay and clock frequency down to deep-cryogenic temperatures.

The powerful advantage of FDSOI over bulk technology is highlighted in this book, which is led by the back biasing capability, and it allows the managing power consumption and performance, which mitigates the thermal effect. These are crucial aspects of cryoelectronics. Involving a combination of heat and mass transfer, the stratification in cryogenic liquid storage systems is a complicated yet inexorable thermodynamic phenomenon. Cryogenic liquids having a very low-boiling point are susceptible to heat entry from the ambient. A large temperature gradient exists between the storage and atmospheric temperature, and is the cause of heat ingress during ground parking. Due to aerodynamic heating (during flight) and space radiation, an overwhelming increase in heat ingress occurs, although these are not as significant as obtained during the coast phase. The propellant tanks receive the insulation of foam that is relatively less effective compared to vacuum or multi-layered insulation (MLI). The temperature of liquid adjacent to the walls is raised due to the mentioned leakage, which induced natural convection currents. Due to buoyancy, the heated liquid starts flowing up and accumulates at the liquid-vapor interface, which creates an axial temperature gradient called thermal stratification. Here, with time the depth of this stratified layer increases. On the other hand, due to vaporization, the tank pressure keeps increasing and it creates the requirement of proper design of venting devices and insulation systems. Thus, for designing the rocket fuel tanks, the thermal stratification is a crucial design criterion for consideration.

On thermal stratification and self-pressurization of a cryogenic storage vessel, a wide variety of experimental and numerical studies have been conducted and reported by researchers across the globe. They studied the effect of rib shape and material thermal conductivity on the development of stratification by considering cylindrical ribbed tanks with a certain percentage (50%) filled volume and rib shapes of rectangular and semi-circular [59]. They observed for rib materials of low thermal conductivity produced lower tank pressurization. Lesser self-pressurization was observed in the semi-circular ribbed tank in comparison with the rectangular ribbed case for the same rib cross-sectional area and locations.

To minimize the thermal stratification in LH2 tanks, some researchers carried out numerical investigations and observed about 30% reduction in the stratification parameter by the application of transverse wall ribs on the inner surface of the cylindrical tank [60]. For the tank with the ribbed inner surface, they noticed a delayed stratification, as well as a lesser natural heat transfer coefficient, in comparison with those output for the smooth wall tank.

Transient natural convection on a vertical ribbed wall was investigated by some researchers [61] and they observed a reduction in convective heat transfer coefficient below the initial rib and enhancement was noticed after that region. Other researchers observed a reduction in heat transfer performance for a natural convective airflow over a heated ribbed plate [62]. To reproduce the thermal field, the Schlieren optical technique was used, and it was found that the induced flow creates thermally inactive regions just upstream and downstream of each protruding element.

Researchers [63] studied the flow behavior of air over a heated wall with single and repeated, two-dimensional, rectangular roughness elements. They did not notice the enhancement of heat transfer in the presence of wall ribs. The volume of fluid (VOF) method was used to investigate the depressurization and thermal stratification behavior of a liquid nitrogen tank with different baffle structures under microgravity conditions [64]. A reduction of up to 54% in pressurization rate was achieved by optimizing the baffle setting. The effect of isogrid on thermal stratification inside propellant tanks was investigated by researchers [65]. They observed that the boundary layer thickness on the wall in a forced-free stream flow was distinctly thicker (150–700%) than the equivalent flat plate boundary layer thickness. Depending upon roughness size and tank conditions, the isogrids can either enhance or suppress the stratification rate compared with smooth tanks. The boundary layer behavior over the propellant tank with mass saving isogrid structures was investigated and observed development of more than 200% thicker velocity boundary layer over isogrid wall than a smooth wall, which led to rapid self-pressurization and enhanced fluid mixing [66].

Thus, the liquid fuels held in liquid form at cryogenic temperature and gas at normal temperatures are used in cryogenic engines. Minimal heat infiltration causes thermal stratification and self-pressurization because the propellants are stored at their boiling temperature or subcooled condition. The state of propellant inside the tank varies due to stratification, and it is essential to keep the propellant properties in a predefined state for restarting the cryogenic engine after the coast phase. Here, cavitation could happen and create destruction of the flight vehicle if the inlet temperature is above the cavitation value. To reduce the stratification phenomenon in a cryogenic storage tank, some investigations are going on to find some effective methods, such as the shape of the inner wall surface of the storage tank plays an essential role in the development of the stratified layer. To predict the rate of self-pressurization in a liquid hydrogen container, an established CFD model could be used. To predict the liquid-vapor interface movement, the Volume of Fluid (VOF) method was used by researchers, whereas the Lee phase change model was adopted for evaporation and condensation calculations.

A detailed study was conducted on a cylindrical storage tank by some researchers with an isogrid and rib structure. Free convection flow of buoyancy-driven support over isogrid structure resulted in significantly different velocity and temperature profiles from a smooth wall case. Thus, the isogrid-type obstruction provided a more significant thermal boundary layer, and those obstructions induced streamline deflection and recirculation zones, which improved the heat transfer to bulk liquid. With an isogrid structure, a larger self-pressurization rate was observed for tanks. A reduction of upward buoyancy flow near the tank surface was obtained with the presence of ribs, whereas with it streamline deflection and recirculation zones were also noticeable. It nullified the effect of the formation of recirculation zones with the increase in the number of ribs. The self-pressurization rate was reduced by about 32.89% with the incorporation of the rib structure in the tank wall.

Thermal stratification is reduced in the presence of roughness elements. The conductivity of ribs on heat transfer performance for the flat vertical and horizontal heated plates has been studied extensively with the variation of spacing to height ratios of transverse ribs, protrusion length, etc. Thus, further investigation is needed to realize the influence of ribs and grid structure on the reduction of stratification on a cryogenic cylindrical tank.

The improvement of manufacturing technology and the expansion of the semiconductor market caused the rise in wafer size and well-integrated semiconductor devices. Due to higher integration, the heat produced in the semiconductor manufacturing process is required to be at a lower level, so the temperature of the supplied cooling medium from the chiller is needed to keep at a lower level. Therefore, the Joule-Thomson cooling cycle is used and, in the process, a mixed refrigerant (MR) is used to produce the cooling medium at a level of −100°C, which is required for the semiconductor process. This technique has gained great attention, where a mixture of refrigerants (MR) is used, and the chiller performance is heavily influenced by the composition and proportions of the refrigerant charged to the chiller system. The application of MR in the cooling system to achieve the required low temperature of −100°C in the semiconductor manufacturing process and the use of different proportions and the compositions of MR and results are represented by some researchers. It was observed that the increased proportion of high-boiling point refrigerant could shorten the time for cooling the process and it keep the device pressure at a lower level. On the other hand, the reduction of the proportion of the mass ratio of the high-boiling point refrigerant enhances the cooling capacity of the refrigerants. However, it was reported that the MR refrigerants could reach the target temperature of the process in a short time and the cooling capacity is high when R290 is used as the high-boiling point refrigerant among R290 and R600a [67].

Author details

Salim Newaz Kazi
Department of Mechanical Engineering, Faculty of Engineering, University of Malaya, Kuala Lumpur, Wilayah Persekutuan Kuala Lumpur, Malaysia

*Address all correspondence to: salimnewaz@um.edu.my

References

[1] Azeotropic/zeotropic refrigerants. Refrigeration Handbook. Sweden: SWEP Company; 2021. https://www.swep.net/refrigerant-handbook/5.-refrigerants/sd1/

[2] Vardelle A et al. The 2016 thermal spray roadmap. Journal of Thermal Spray Technology. 2016;**25**(8):1376-1440

[3] Heberlein J, Murphy AB. Thermal plasma waste treatment. Journal of Physics D: Applied Physics. 2008; **41**(5):053001

[4] Lee CG, Kanarik KJ, Gottscho RA. The grand challenges of plasma etching: A manufacturing perspective. Journal of Physics D: Applied Physics. 2014;**47**(27): 273001

[5] Von Woedtke T, Metelmann HR, Weltmann KD. Clinical plasma medicine: State and perspectives of in vivo application of cold atmospheric plasma. Contributions to Plasma Physics. 2014;**54**(2):104-117

[6] Agarwal P, Girshick SL. Numerical modeling of an RF argon–silane plasma with dust particle nucleation and growth. Plasma Chemistry and Plasma Processing. 2014;**34**(3):489-503

[7] Chen Q, Li J, Li Y. A review of plasma–liquid interactions for nanomaterial synthesis. Journal of Physics D: Applied Physics. 2015;**48**(42):424005

[8] Adamovich I et al. The 2017 plasma roadmap: Low temperature plasma science and technology. Journal of Physics D: Applied Physics. 2017; **50**(32):323001

[9] Paik JK, Lee DH, Noh SH, Park DK, Ringsberg JW. Full-scale collapse testing of a steel stiffened plate structure under cyclic axial-compressive loading. In: Structures. Netherlands: Elsevier; Aug 2020;**26**:996-1009

[10] JK P. Ultimate Limit State Analysis and Design of Plated Structures. Chichester, UK: John Wiley & Sons; 2018

[11] Paik JK. Advanced Structural Safety Studies: With Extreme Conditions and Accidents. Vol. 37. USA: Springer; 2019

[12] Brockenbrough RL, Johnston BG. Steel Design Manual. USA: United States Steel Corporation; 1981. ADUSS 27-3400-04, January

[13] Dowling N. Mechanical Behavior of Materials. Fourth ed. New York, USA: PEARSON; 2012

[14] Yoo S-W et al. Temperature and strain rate dependent constitutive model of TRIP steels for low-temperature applications. Computational Materials Science. 2011;**50**(7):2014-2027

[15] Choung J, Nam W, Lee J-Y. Dynamic hardening behaviors of various marine structural steels considering dependencies on strain rate and temperature. Marine Structures. 2013;**32**:49-67

[16] Chen L et al. Probabilistic and constitutive models for ductile-to-brittle transition in steels: A competition between cleavage and ductile fracture. Journal of the Mechanics and Physics of Solids. 2020;**135**:103809

[17] Cerik BC, Park S-J, Choung J. Use of localized necking and fracture as a failure criterion in ship collision analysis. Marine Structures. 2020; **73**:102787

[18] Paik JK et al. Full-scale collapse testing of a steel stiffened plate structure under axial-compressive loading triggered by brittle fracture at cryogenic condition. Ships and Offshore Structures. 2020;**15**(sup.1):S29-S45

[19] Paik JK et al. Full-scale collapse testing of a steel stiffened plate structure under axial-compressive loading at a temperature of– 80° C. Ships and Offshore Structures. 2021; **16**(3):255-270

[20] Bolaji BO. Theoretical assessment of new low global warming potential refrigerant mixtures as eco-friendly alternatives in domestic refrigeration systems. Scientific African. 2020; **10**:e00632

[21] Banjo SO, Bolaji BO, Ajayi OO, Olufemi BP, Osagie I, Onokwai AO. Performance enhancement using appropriate mass charge of R600a in a developed domestic refrigerator. In: IOP Conference Series: Earth and Environmental Science. Bristol, England: IOP Publishing; 2019;**331**(1):012025

[22] Desai P. Modern Refrigeration and Air Conditioning for Engineers Part II. Delhi: Khana Publishers; 2009

[23] Kang GH, Na SI, Kim MS. Reasonable Performance Comparison for a Refrigeration System Using Different Refrigerants: Case with Propane and Isobutane Mixtures with Several Compositions. International Journal of Refrigeration. Elsevier; 2021

[24] McMullan JT. Refrigeration and the environment—Issues and strategies for the future. International Journal of Refrigeration. 2002;**25**(1):89-99

[25] Abas N et al. Natural and synthetic refrigerants, global warming: A review. Renewable and Sustainable Energy Reviews. 2018;**90**:557-569

[26] Cengel YA, Boles MA, Kanoglu M. Thermodynamics: An Engineering Approach. Vol. 5. New York: McGraw-Hill; 2011

[27] Banjo SO, Bolaji BO, Osagie I, Fayomi OSI, Fakehinde OB, Olayiwola PS, et al. Experimental analysis of the performance characteristic of an eco-friendly HC600a as a retrofitting refrigerant in a thermal system. In: Journal of Physics: Conference Series. Bristol, England: IOP Publishing; 2019;**1378**(4):042033

[28] Applied Thermodynamics: For Engineering Technologists. Fifth Edition. In : Eastop TD, McConkey A. Burnt Mill, Harlow, Longman Higher Education; 1993

[29] Bolaji BO. Selection of environment-friendly refrigerants and the current alternatives in vapour compression refrigeration systems. Top Research Group Journal of Science and Management; 2011;**1**(1):22-26. http://repository.fuoye.edu.ng/handle/123456789/1409

[30] Tashtoush B, Tahat M, Shudeifat M. Experimental study of new refrigerant mixtures to replace R12 in domestic refrigerators. Applied Thermal Engineering. 2002;**22**(5):495-506

[31] Wongwises S, Chimres N. Experimental study of hydrocarbon mixtures to replace HFC-134a in a domestic refrigerator. Energy Conversion and Management. 2005;**46**(1):85-100

[32] Sekhar SJ, Lal DM. HFC134a/HC600a/HC290 mixture a retrofit for CFC12 systems. International Journal of Refrigeration. 2005;**28**(5):735-743

[33] Raghunatha Reddy DV, Bhramara P, Govindarajulu K. Performance analysis of domestic refrigerator using hydrocarbon refrigerant mixtures with ANN and Fuzzy logic system. In: Numerical Heat Transfer and Fluid Flow. Singapore: Springer; 2019. pp. 113-121

[34] United Nations Environment Programme. Ozone Secretariat. Handbook for the International Treaties for the Protection of the Ozone Layer:

The Vienna Convention (1985), the Montreal Protocol (1987). UNEP/ Earthprint; 2003

[35] Adekunle A et al. Assessment of carbon emissions for the construction of buildings using life cycle analysis: Case study of Lagos state. International Journal of Engineering Research and Advance Technology. 2020;**6**(8):1-11

[36] Peyyala A, Sudheer NVVS. Experimental investigation of COP using hydro carbon refrigerant in a domestic refrigerator. In: IOP Conference Series: Materials Science and Engineering. Bristol, England: IOP Publishing; 2017;**225**(1):012236

[37] Sanket B. Substitution of non-ecofriendly refrigerants by hydrocarbon refrigerants: A review. International Journal of Advance in Science Engineering and Technology Aurangabad. 2016;**2**:122-125

[38] Kundu A, Kumar R, Gupta A. Performance comparison of zeotropic and azeotropic refrigerants in evaporation through inclined tubes. Procedia Engineering. 2014;**90**:452-458

[39] Qureshi MA, Bhatt S. Comparative analysis of COP using R134a & R600a refrigerant in domestic refrigerator at steady state condition. International Journal of Science and Research (IJSR). 2014;**3**(12):935-939

[40] Ahamed JU et al. An analysis of energy, exergy, and sustainable development of a vapor compression refrigeration system using hydrocarbon. International Journal of Green Energy. 2012;**9**(7):702-717

[41] Sattar M, Saidur R, Masjuki HH. Performance investigation of domestic refrigerator using pure hydrocarbons and blends of hydrocarbons as refrigerants. World Academy of Science, Engineering and Technology, Paris/ France; 2007;**5**:223-228

[42] Harby K. Hydrocarbons and their mixtures as alternatives to environmental unfriendly halogenated refrigerants: An updated overview. Renewable and Sustainable Energy Reviews. 2017;**73**:1247-1264

[43] UN Environment. United Nations Environment Programme, ASHRAE. FACTSHEET 1, Update on New Refrigerants Designations and Safety Classifications, Standard; 2020;34

[44] Balestra F, Ghibaudo G. Device and Circuit Cryogenic Operation for Low Temperature Electronics. Norwell, MA: Kluwer Academic Publishers; 2001. pp. 3-13

[45] Gutierrez-D EA, Deen J, Claeys C. Low Temperature Electronics: Physics, Devices, Circuits, and Applications. Elsevier; 2000

[46] Kirschman R. Cold electronics: An overview. Cryogenics. 1985;**25**(3): 115-122

[47] Hornibrook J et al. Cryogenic control architecture for large-scale quantum computing. Physical Review Applied. 2015;**3**(2):024010

[48] Charbon E, Sebastiano F, Vladimirescu A, Homulle H, Visser S, Song L, et al. Cryo-CMOS for quantum computing. In: 2016 IEEE International Electron Devices Meeting (IEDM). IEEE; 2016. pp. 13-15

[49] Incandela RM, Song L, Homulle HA, Sebastiano F, Charbon E, Vladimirescu A. Nanometer CMOS characterization and compact modeling at deep-cryogenic temperatures. In: 2017 47th European Solid-State Device Research Conference (ESSDERC). IEEE; September 2017. pp. 58-61

[50] Beckers A, Jazaeri F, Ruffino A, Bruschini C, Baschirotto A, Enz C. Cryogenic characterization of 28 nm bulk CMOS technology for quantum computing. In: 2017 47th European

Solid-State Device Research Conference (ESSDERC). IEEE; September 2017. pp. 62-65

[51] Hart M, Evans A. Rapid thermal processing in semiconductor technology. Semiconductor Science and Technology. 1988;**3**(5):421

[52] Balestra F, Ghibaudo G. Physics and performance of nanoscale semiconductor devices at cryogenic temperatures. Semiconductor Science and Technology. 2017;**32**(2):023002

[53] Shin M, Shi M, Mouis M, Cros A, Josse E, Kim GT et al. Low temperature characterization of 14nm FDSOI CMOS devices. In: 2014 11th International Workshop on Low Temperature Electronics (WOLTE). IEEE; July 2014. pp. 29-32

[54] Bohuslavskyi H, Barraud S, Cassé M, Barrai V, Bertrand B, Hutin L et al. 28nm fully-depleted SOI technology: Cryogenic control electronics for quantum computing. In: 2017 Silicon Nanoelectronics Workshop (SNW). IEEE; June 2017. pp. 143-144

[55] Le Guevel L, Billiot G, Jehl X, De Franceschi S, Zurita M, Thonnart Y et al. 19.2 a 110mK 295μW 28nm FDSOI CMOS quantum integrated circuit with a 2.8 GHz excitation and nA current sensing of an on-Chip double quantum dot. In: 2020 IEEE International Solid-State Circuits Conference-(ISSCC). IEEE; February 2020. pp. 306-308

[56] Noel J-P et al. Multi-V_T UTBB FDSOI device architectures for low-power CMOS circuit. IEEE Transactions on Electron Devices. 2011;**58**(8): 2473-2482

[57] Sesnic SS, Craig GR. Thermal effects in JFET and MOSFET devices at cryogenic temperatures. IEEE Transactions on Electron Devices. 1972;**19**(8):933-942

[58] Foty D, Titcomb S. Thermal effects in n-channel enhancement MOSFET's operated at cryogenic temperatures. IEEE Transactions on Electron Devices. 1987;**34**(1):107-113

[59] Fu J, Sunden B, Chen X. Influence of wall ribs on the thermal stratification and self-pressurization in a cryogenic liquid tank. Applied Thermal Engineering. 2014;**73**(2):1421-1431

[60] Khurana TK et al. Thermal stratification in ribbed liquid hydrogen storage tanks. International Journal of Hydrogen Energy. 2006;**31**(15): 2299-2309

[61] Polidori G, Padet J. Transient free convection flow on a vertical surface with an array of large-scale roughness elements. Experimental Thermal and Fluid Science. 2003;**27**(3):251-260

[62] Tanda G. Natural convection heat transfer in vertical channels with and without transverse square ribs. International Journal of Heat and Mass Transfer. 1997;**40**(9):2173-2185

[63] Hadim A, Ramot M. Natural Convection in an Enclosure with Discrete Heat Sources on the Vertical Walls. ASME, Journal of Electronic Packaging. 1993. DOI: 10.1115/1.290935

[64] Zuo Z, Jiang W, Huang Y. Effect of baffles on pressurization and thermal stratification in a LN2 tank under micro-gravity. arXiv preprint. 2018;arXiv:1807.01547. [physics.flu-dyn], DOI: 10.1016/j.cryogenics.2018.10.017.

[65] Oliveira J et al. Effect of isogrid roughness on thermal stratification. In: 43rd AIAA/ASME/SAE/ASEE Joint Propulsion Conference & Exhibit Cincinnati, OH;. 2007. p. 5497. DOI: 10.2514/6.2007-5497

[66] Faure JM et al. Effect of isogrid-type obstructions on thermal stratification in

upper-stage rocket propellant tanks. Journal of Spacecraft and Rockets. 2014;**51**(5):1587-1602

[67] Yoon JI, Son CH, Seol SH, Yoon JH. Ultra-low temperature chillers for semiconductor manufacturing process. In: Advancements Towards Sustainability of HVAC & R Systems. London, UK: IntechOpen; 2021

Chapter 2

Low Temperature Characterization and Modeling of FDSOI Transistors for Cryo CMOS Applications

Mikaël Cassé and Gérard Ghibaudo

Abstract

The wide range of cryogenic applications, such as spatial, high performance computing or high-energy physics, has boosted the investigation of CMOS technology performance down to cryogenic temperatures. In particular, the readout electronics of quantum computers operating at low temperature requires larger bandwidth than spatial applications, so that advanced CMOS node has to be considered. FDSOI technology appears as a valuable solution for co-integration between qubits and consistent engineering of control and read-out. However, there is still lack of reports on literature concerning advanced CMOS nodes behavior at deep cryogenic operation, from devices electrostatics to mismatch and self-heating, all requested for the development of robust design tools. For these reasons, this chapter presents a review of electrical characterization and modeling results recently obtained on ultra-thin film FDSOI MOSFETs down to 4.2 K.

Keywords: Cryogenic CMOS, FDSOI, MOSFET, characterization, modeling

1. Introduction

First MOSFET measurements at liquid Helium temperature have been reported as soon as in late 1960s [1–3], leading to some remarkable discoveries like the integer quantum Hall effect [4]. Since then, many works have been published on the electrical characteristics of MOSFETS down to 4.2 K [5–7]. The interest of operating electronic circuits at cryogenic temperatures has been demonstrated a few decades ago, and relies on the performance improvement and/or on the necessity to have electronics in cryogenic environment [5–7]. With the emerging field of quantum computing, for which read-out and control electronics of the quantum bits (qubits) is required in the proximity of the qubit itself, the study of CMOS devices at low and very low temperature, well below 100 K, has received a renewed attention [8–10]. In particular, qubit control requires high-frequency and large-bandwidth signals, as well as low-power electronics to be compatible with the cooling power of modern refrigerators [11–14]. Circuits fabricated from advanced nodes CMOS are good candidates to fulfill the specifications for quantum computing applications [15–18].

Key advantages of operating at low temperatures include the better electrical performance of MOSFETs, with higher carrier drift velocity and so higher on-state drain current and transconductance, steeper subthreshold slope, lower leakage current [6, 19]. Some works have studied bulk MOSFETs operation at cryogenic temperature emphasizing in particular kink behavior and freeze-out effects in those devices [7, 15, 20–24]. Recently, outstanding characteristics have been demonstrated at 4.2 K on advanced CMOS technologies [19, 25–27], in particular for Fully Depleted Silicon-On-Insulator (FDSOI) [28–32]. Ultrathin film FDSOI devices (with typically silicon thickness less than 10 nm) are immune to kink effects [33], and freeze-out has finally little impact on the DC characteristics of MOSFETs in advanced technologies [34]. Apart from the performance itself of the circuits at these low temperatures, and the figures of merit for analog or digital applications, specific attention to power dissipation has to be brought as well, as the available cooling power is limited in cryostat, and depends of the different cooling stages (typically $\approx$1 W at 4 K and less than 1 mW below 100mk) [11].

In that context, FDSOI technology offers a significant advantage over other available technologies, as it allows designing low power electronics, threshold voltage tunability thanks to its back bias ability, and offers low variability due to the undoped channel [35]. Extensive electrical characterization of advanced CMOS devices at deep cryogenic operation, including device electrostatics, carrier transport, mismatch and variability, or self-heating, is thus seriously needed.

Numerical issues appears with the modeling and simulation of MOSFETs at cryogenic and deep–cryogenic temperatures, in particular due to energy k_BT approaching zero in equations and the extremely small intrinsic carrier density [34, 36]. Besides these difficulties, accurate models must correctly include, among other things, the temperature dependence of the main electrical parameters, such as carrier mobility, saturation velocity, threshold voltage, … , as well as thermal effects [37, 38]. On the other hand, new physical phenomena appear as the device temperature decreases that need to be characterized and properly modeled [19].

Because these aspects are essential for the development of compact models and robust design tools, this chapter presents a review of recent results obtained on 28 nm FDSOI transistors operated down to deep cryogenic temperatures. More specifically, we first discuss in Section 2 the major device electrical properties in terms of transfer characteristics and MOSFET parameters versus temperature. Then, we describe in Section 3 the self-heating phenomena, which could alter the FDSOI device performances. The matching and variability properties of scaled transistors limiting the analog applications are then addressed in Section 4. The development of compact model necessary for FDSOI circuit design at deep cryogenic temperatures is presented in Section 5. Finally, in Section 6, we illustrate the operation of elementary circuits at very low temperatures regarding inverter delay and oscillator frequency.

2. Cryogenic FDSOI device operation

In this section, we present the measurement of the main electrical properties of FDSOI devices operating down to 4.2 K, such as the capacitance and charge control characteristics, the drain current $I_d(V_g)$ transfer curves as well as the main MOSFET parameters (threshold voltage V_{th}, subthreshold swing, mobility).

2.1 Devices under test

The measurements were performed on 28 nm FDSOI MOSFETs with silicon film thickness t_{si} = 7 nm and buried oxide (BOX) thickness t_{BOX} = 25 nm from STMicroelectronics. NMOS and PMOS transistors were processed from (100) handle substrate, with <100> – oriented channel, and a high-κ/metal gate Gate-First architecture (**Figure 1**) [39]. Regular-V_{th} (RVT) and low-V_{th} (LVT) transistors are available through a doped back plane (NWELL or PWELL, with typically $N_{A,D}$ = 10^{18} cm^{-3}) below the BOX. Thin (GO1, with equivalent oxide thickness EOT = 1.1 nm) and thick oxide (GO2, EOT = 3.2 nm) devices have been characterized using a cryogenic probe station down to 4.2 K.

2.2 Capacitance and charge control

The electrostatic charge control of FDSOI devices has been characterized by split C-V measurements with a conventional LCR meter. To this end, the gate-to-channel capacitance C_{gc} = dQ_i/dV_g, with Q_i the inversion charge in the channel, has been measured at 500 kHz frequency on large area N and P MOS devices as a function of the front gate voltage V_g with body bias V_b = 0 V for several temperatures down 4.2 K (**Figure 2**). As can be seen, the $C_{gc}(V_g)$ curves are almost temperature independent above threshold, whereas a strong improvement of the turn-on behavior is obtained at low temperature, related to the subthreshold slope increase. These characteristics have been well reproduced by Poisson-Schrodinger simulations (see **Section 5.1**), providing precise extraction of front oxide EOT values for GO1 and GO2 transistors [41].

The influence of the AC level (V_{osc}) of the LCR meter oscillator used during C_{gc} measurements at 4.2 K has been studied and is reported in **Figure 3a**. Indeed, due to the strong non linearity of the $Q_i(V_g)$ curves in subthreshold region at very low temperature, the turn-on behavior of the $C_{gc}(V_g)$ curve below threshold is not well captured for a too large AC level (here 40 mV, currently used at T = 300 K). However, for an AC level of 1 mV, getting closer to the thermal voltage k_BT/q at

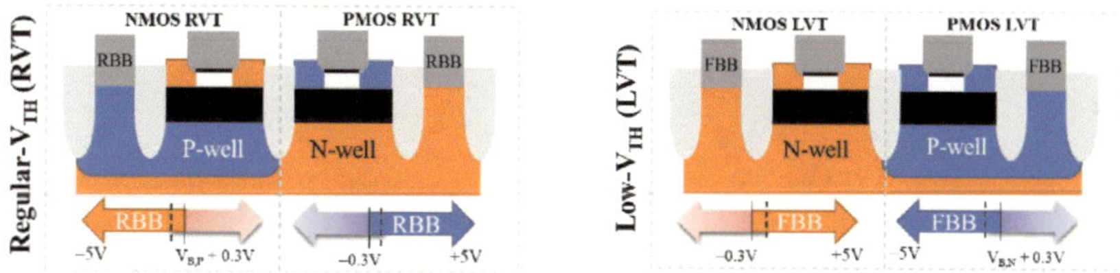

Figure 1.
Schematics of 28 nm FDSOI N- and PMOSFETs with regular-V_{TH} (RVT) and low-V_{TH} (LVT) flavors. Forward and reverse back biases (FBB and RBB) can be applied depending on the doping of the back plane.

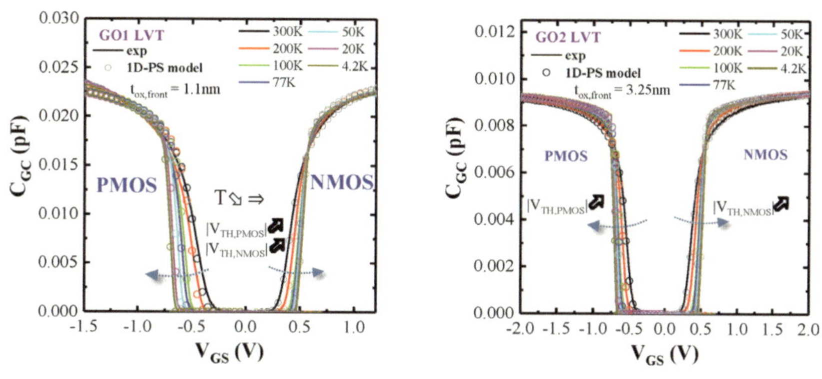

Figure 2.
$C_{gc}(V_g)$ characteristics (solid lines) for N- and PMOS GO1 and GO2 devices from 300 K down to 4.2 K, at V_B = 0 V. the $C_{gc}(V_g)$ 1D-PS modeling is shown in symbols (frequency = 1 MHz, AC level = 40 mV, W = L = 9 μm). After Cardoso et al. *[40].*

4.2 K, where k_B is the Boltzmann constant and q the magnitude of the electron charge, the turn-on behavior of $C_{gc}(V_g)$ below threshold is well accounted for. These results can be well modeled by integrating the ideal $C_{gc}(V_g)$ curve over one period of the AC signal, providing the measured capacitance $C_{gc,meas}$ as follows [42]:

$$C_{gc,meas}(V_g) = \frac{1}{Tp}\int_0^{T_P} C_{gc}(V_g + \delta V_g(t))dt \quad (1)$$

where $\delta V_g(t) = V_{osc}.\sin(2\pi t/T_p)$ is the AC signal of period T_p (**Figure 3b**).

2.3 Drain current characteristics, threshold voltage and subthreshold slope

The $I_d(V_g)$ transfer characteristics of same devices have been measured in linear region (V_d = 50 mV) for various temperatures and are shown in **Figure 4**. As usually observed in cryo-electronics for bulk CMOS devices [7], the drain current above threshold is highly increased due to mobility improvement of both electrons and holes, resulting from the suppression of phonon scattering. Similarly, the turn-on behavior of the curves below threshold is greatly improved as the temperature is lowered.

The threshold voltage V_{th} of the devices has been extracted by the constant current method (*i.e.* V_g for which $I_d = 10^{-7} \times W/L$) and typical variations with

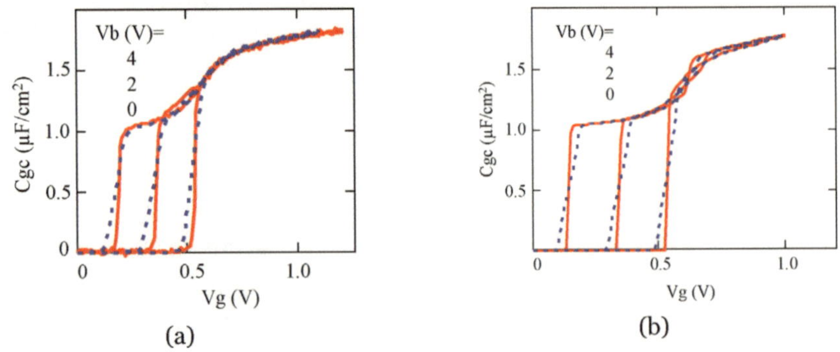

Figure 3.
Experimental (a) and modeled (b) $C_{gc}(V_g)$ characteristics for NMOS GO1 devices (W = L = 10 μm) at 4.2 K for two AC levels: 40 mV (red solid lines) and 1 mV (blue dashed lines).

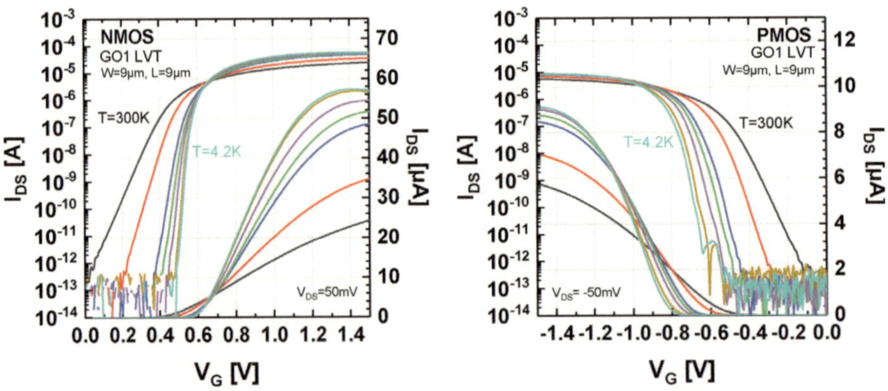

Figure 4.
$I_d(V_g)$ characteristics for GO1 N and P MOS devices for various temperatures obtained in linear region (V_d = 50 mV).

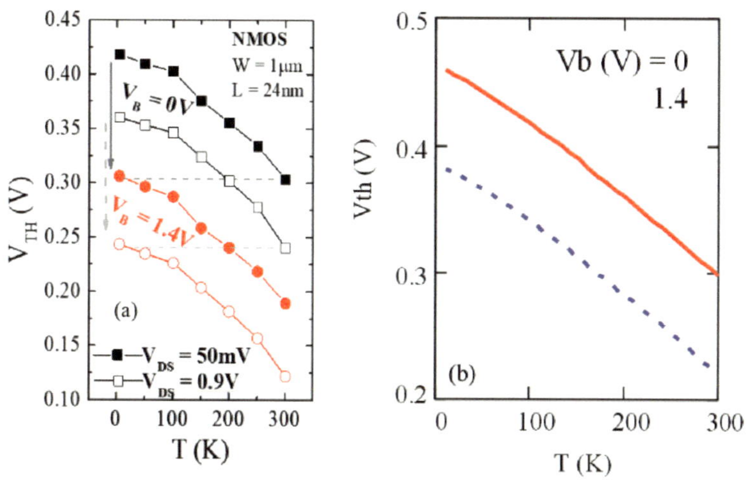

Figure 5.
(a) Experimental V_{th} extracted on NMOS GO1 transistor (W = 1 μm, L = 24 nm) as a function of T at V_{DS} = 50 mV and 0.9 V, and at V_b = 0 V and 1.4 V. (b) Modeled V_{th} vs. T for V_{ds} = 50 mV and V_b = 0 and 1.4 V. After Cardoso et al. [43].

temperature are shown in **Figure 5a**. As in bulk MOS devices [7], V_{th} increases as the temperature is reduced, here with sensitivity around 0.7 to 1 mV/K. It should be mentioned that in FDSOI devices with undoped film as in our case, the V_{th} variation with T is not explained by the temperature dependence of the Fermi level in the silicon film as for bulk MOS devices [44]. Actually, a simple model for V_{th} read by constant current method can be derived assuming a single subband for the inversion layer with a critical inversion charge density n_{th} as:

$$V_{th} = V_{sth} + \frac{q.n_{th}}{C_{ox}} + \frac{C_b.(V_{sth} - V_b)}{C_{ox}} \tag{2}$$

with $V_{sth} = V_0 + \frac{k_B T}{q}.ln\left(e^{\frac{n_{th}}{k_B T.A_{2D}}} - 1\right)$ being a threshold surface potential associated with a given constant inversion charge density n_{th} (here $10^{10}/cm^2$), and where V_0 is a constant, A_{2D} the 2D subband density of states, C_{ox} and C_{box} respectively the front gate oxide and the buried oxide capacitance and $C_b = C_{box}.C_{si}/(C_{si} + C_{box})$ the body to front channel coupling capacitance. As can be seen from **Figure 5b**, a good qualitative agreement between model and experiment can be achieved with Eq. (2).

An important feature of FDSOI devices is the strong V_{th} control allowed by the back bias, which is not possible in FinFET and NW architectures, and very limited in bulk MOS devices [7], especially in forward biasing. Typical dependence of V_{th} with back bias are illustrated in **Figure 6** for both P and N MOS FDSOI devices of various flavors and gate oxide thicknesses (GO1 and GO2), at T = 4.2 K and T = 300 K. As can be seen from this figure, it appears that the threshold voltage control with back biasing ($\Delta V_{th}/\Delta V_b$) is insensitive to temperature down to cryogenic conditions, and that V_{th} can be decreased to values close to zero volt. Interestingly, this makes it possible to operate the FDSOI devices at deep cryogenic temperatures with very small supply voltage (≈0.1–0.2 V), enabling low power dissipation.

Another important parameter in FET operation is the so called subthreshold slope, $S = dln(I_d)/dV_g$, or its inverse the subthreshold swing SS, which characterizes the turn-on efficiency of the MOSFET below threshold. Typical subthreshold swing SS (mV/dec) variations with drain current in weak inversion region are shown in

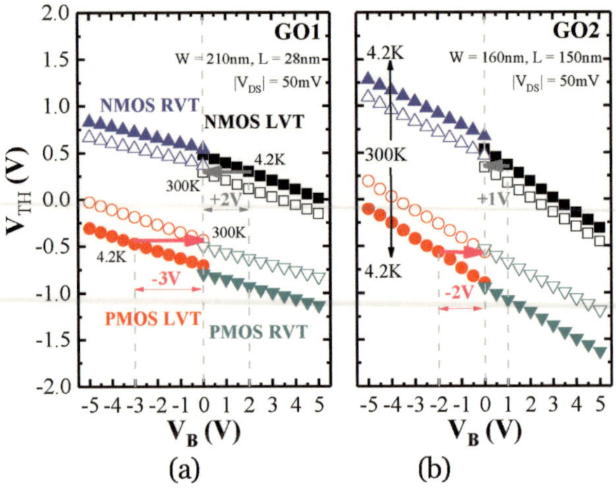

Figure 6.
Measurements of V_{th} vs. V_b for N- and P-type, RVT and LVT, GO1 (a) and GO2 (b) MOSFETs, at 300 K and 4.2 K, V_{DS} = 50 mV. As T is decreased, V_b can be used to shift V_{th} back to its value at room temperature. After Cardoso et al. [40].

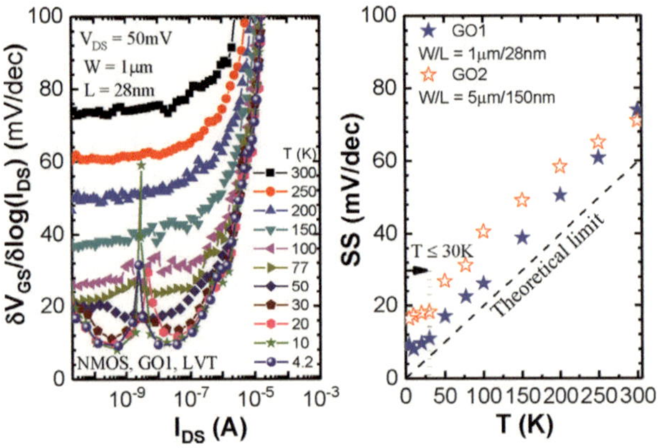

Figure 7.
Extracted subthreshold current vs. *I_d (a) and SS vs. T (b) for NMOS LVT from 300 K to 4.2 K. After Cardoso et al. [45].*

Figure 7a, revealing a plateau from which an average subthreshold swing can be extracted and plotted versus temperature (**Figure 7b**). Indeed, the subthreshold swing SS is varying linearly with temperature down to 25-30 K before plateauing around 10-20 mV/decade at deep cryogenic temperatures. The SS(T) linear behavior is usual for all FET devices and simply related to the Maxwell-Boltzmann statistics prevailing in weak inversion where SS = $kT/q.(C_{ox} + C_b + C_{it})/C_{ox}$, C_{it} being the interface trap density capacitance [7]. The SS(T) plateau is generally attributed to the presence of an exponential tail of subband states, likely due to potential-fluctuations-induced disorder [46–48] and that minimizes the drain current turn-on efficiency at deep cryogenic temperatures.

2.4 Carrier mobility

Finally, the effective carrier mobility μ_{eff} is investigated as being a driving parameter of MOSFET in linear region. In **Figure 8a** and **b** are illustrated typical mobility variations with inversion charge N_{inv} as obtained by split C-V method in

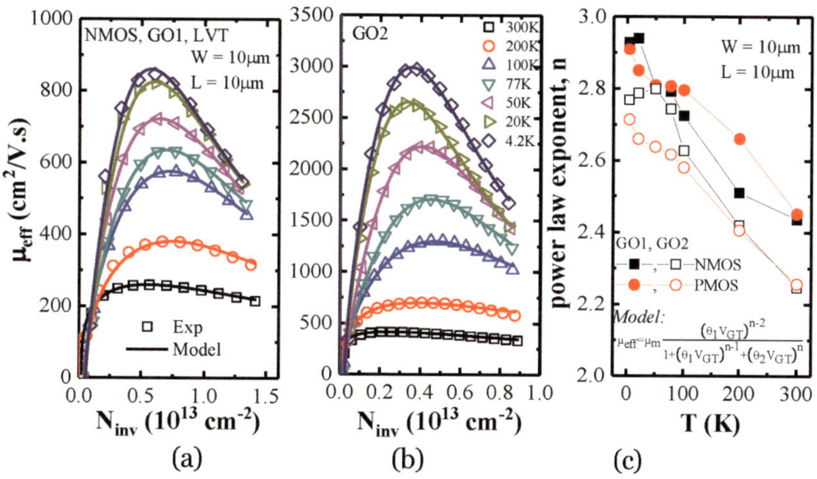

Figure 8.
Experiments and analytical model of μ_{eff} vs. N_{inv} for NMOS GO1 (a) and GO2 (b), varying T. power law exponent (c) vs. T for N and PMOS. After Cardoso et al. *[49].*

such FDSOI MOS devices for various temperatures. As can be seen, there is a strong improvement (up to 10 times) of the maximum mobility with temperature lowering due to phonon scattering reduction [7]. As already found for bulk Si MOSFET [7], the effective mobility exhibits a bell-shaped behavior with inversion charge at low temperature, where the mobility is limited by combined Coulomb and surface roughness scattering processes. As also shown in **Figure 8**, the mobility can be well fitted by an empirical model inspired from bulk MOSFET results and written as:

$$\mu_{eff} = \mu_m \cdot \frac{\left(\theta_1 \cdot \frac{Q_i}{C_{ox}}\right)^{n-2}}{1 + \left(\theta_1 \cdot \frac{Q_i}{C_{ox}}\right)^{n-1} + \left(\theta_2 \cdot \frac{Q_i}{C_{ox}}\right)^{n}} \tag{3}$$

where μ_m stands for an amplitude mobility value close to the maximum one, θ_1 and θ_2 are the first and second order attenuation coefficients and n is a power law exponent varying between ≈2 and ≈3 as the temperature is changed from 300 K down to 4.2 K, as illustrated in **Figure 8c**. It should be noted that this mobility law *vs*. inversion charge will be useful for compact modeling purpose (see **Section 5**).

As was already mentioned, a specific feature of FDSOI devices is their operation in forward back biasing condition, enabling a significant lowering of the threshold voltage as illustrated in **Figure 9a** for T = 4.2 K. Interestingly, for sufficiently large V_b, the drain current measured at low V_d and very low temperatures (here T = 4.2 K) is increasing above back channel threshold before to decrease significantly and then to increase again well above front channel threshold. Actually, this decrease of the drain current just happens when the front channel is opening and has been attributed to a reduction of the mobility due to remote inter-subband scattering (IS) as well explained in [50]. To better understand this behavior, we have computed the drain current of the back channel after subtraction of the front channel component, taken as being the one in absence of back channel formation i.e. when V_b = 0 V (see **Figure 9b**). This assumption has been validated by Poisson-Schrodinger simulation (not shown here). Doing the same with $C_{gc}(V_g)$ characteristics for various V_b's, the inversion charge in the back channel has also been computed after integration of capacitance vs. V_g as is usual in split C-V technique (**Figure 10a**). As a result, note that the back channel charge is plateauing after the

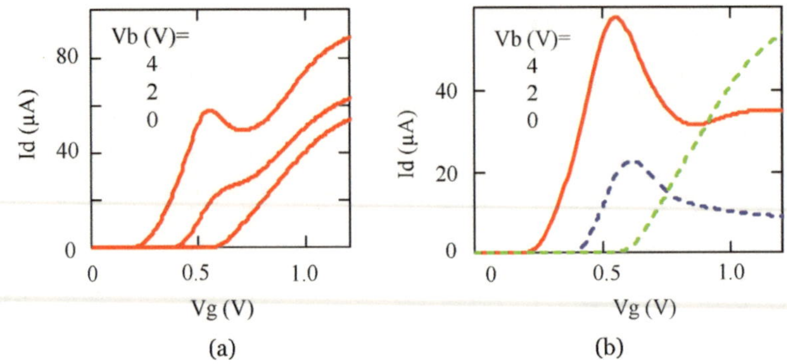

Figure 9.
a) $I_d(V_g)$ characteristics at 4.2 K for various V_b (= 0, 2 V, 4 V) and b) Back channel $I_d(V_g)$ curves after subtraction of front channel component also shown in green dashed line.

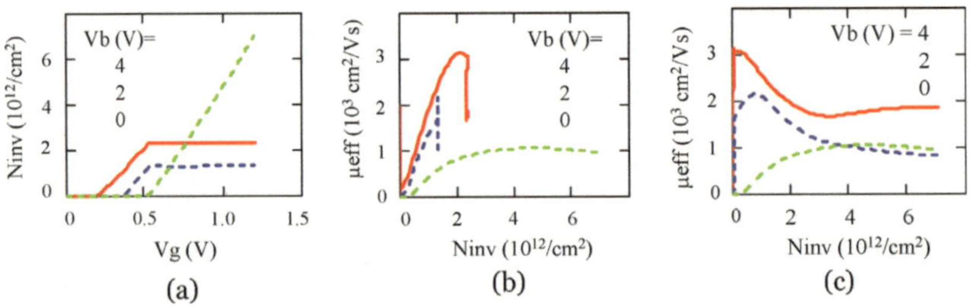

Figure 10.
a) N_{inv} vs. V_g for Back channel for V_b = 2 V, 4 V. Green curve shows $N_{inv}(V_g)$ for front channel at V_b = 0. b) Back channel μ_{eff} vs. back channel N_{inv} and c) Back channel μ_{eff} vs. front channel N_{inv} for V_b = 2 V, 4 V. Green curves show front channel μ_{eff} vs. front channel N_{inv} at V_b = 0 V.

opening of front channel. The effective mobility in the back channel has been computed and plotted versus inversion charge density in the back channel or versus the front one as shown in **Figure 10b** and **c**. As can be seen, μ_{eff} first increases with the back channel inversion charge density before to decrease as the back channel charge saturates (**Figure 10b**). Instead, μ_{eff} in back channel decreases with the front channel inversion charge, which clearly indicates that the opening of the front channel is responsible for the back channel mobility decrease. This is precisely the signature of remote inter-subband scattering, which happens when carriers in the back interface 2D subband can interact with the front interface 2D subband. In this situation, some carriers at the back interface can experience scattering mechanisms in the front interface due to the overlap of the back and front subband wave functions. It should be mentioned that this phenomenon of inter-subband scattering is canceling out when the temperature is increased (T > 50 K) due to thermal broadening as well as when the drain voltage is increased due to the averaging over the channel of the conductance by integration over space [50].

3. Self-heating phenomena

In FDSOI devices or multi-gate field effect transistors like FinFETs and nanowire FETs, low thermally conductive materials such as the buried oxide (BOX) or the thin Si layer constituting the channel hinder the dissipation of the heat generated in the drain side. Consequently, the channel temperature can significantly rise when the device is in ON operation. This self-heating effect (SHE) can in turn severely affect the device performance, by reducing the carrier mobility, shifting the threshold voltage [51] or degrading the device reliability [52, 53], with implications to IC design. SHE has been widely studied for room temperature operation of circuits [54]. The thermal effects play a more fundamental role in cryogenic electronics – operating at various temperature stages with different available cooling powers –, as the temperature increase due to SH can be of the same order or even higher than the ambient temperature [55]. Furthermore, at very low temperature (well below 1 K), the cooling power drops down drastically (typically, 1 W at 1 K, 1 mW at 100mK) and thermal management thus becomes an additional constraint.

In this regard, the study of self-heating effects at cryogenic temperatures provides valuable information for performance optimization. In addition, to be accurate at cryogenic temperatures, models must take into account these thermal effects, as the device temperature can deviate significantly from the ambient one.

3.1 Self-heating characterization technique

The experimental evaluation of self-heating was performed by using the conventional DC technique based on gate resistance thermometry [56]. In this method, the gate dielectric layer is thin enough to assume that the temperature of the channel is equal to that of the gate electrode. Inset of **Figure 11** shows the typical 2-terminal gate structure that we used to measure the gate resistance R_G. R_G is measured between two contacts G1 and G2 using an LCR-meter. By varying the ambient temperature T_{amb} from 4.2 K up to 300 K, we record the change in the electrical gate resistance as a function of the input power $P = I_{DS} \times V_{DS}$. The temperature increase ΔT is deduced from R_G values at zero power (and so without SHE). Then the differential thermal resistance, $R_{TH}^* = \partial\Delta T/\partial P|_{Tamb}$ can be defined. This differential thermal resistance relates the change of ΔT due to a change in power dissipation P at a given T_{amb} [55].

3.2 Study of thermal resistance

In **Figure 11** we have plotted the differential thermal resistance measured on an ultrathin film FDSOI transistor (t_{Si} = 11 nm) as a function of the device temperature

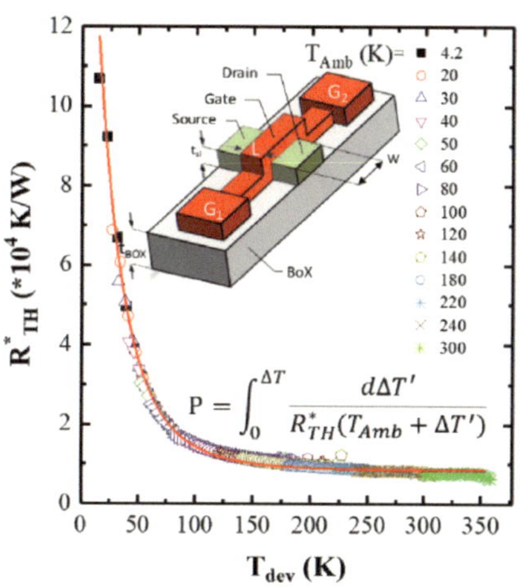

Figure 11.
(symbols) Differential thermal resistance R_{TH}^* measured as a function of the device temperature $T_{device} = T_{amb} + \Delta T$ from 450 K down to 4.2 K, with a corresponding numerical fitting curve (line).

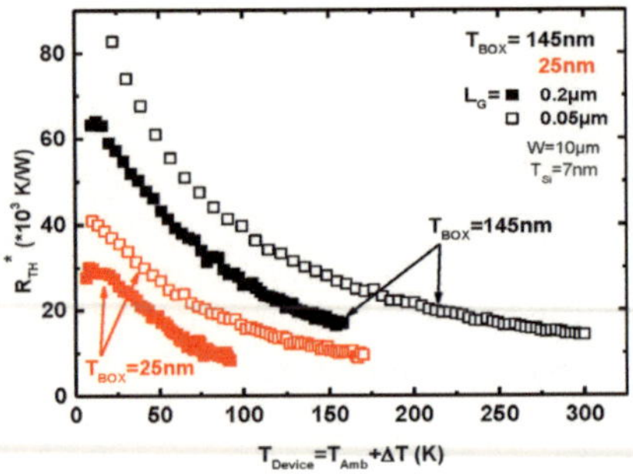

Figure 12.
Thermal resistance R_{TH}^* versus device temperature, for wide and ultrathin FDSOI MOSFETs. After Triantopoulos et al. *[55].*

defined as $T_{device} = T_{amb} + \Delta T$. All the R_{TH}^* data acquired for various ambient temperatures and dissipated power values merge into a single R_{TH}^* versus T_{device} curve, which thus provides a complete description of the temperature dependence of the thermal resistance for a given device. The thermal resistance depends mainly on the device geometry W and L, as well as on the BOX thickness, but not significantly on the Si film thickness in the 7 nm to 24 nm range typical of FDSOI devices (**Figure 12**) [55].

Our results show that in thin film devices, the thermal resistance R_{TH}^* of the device is strongly temperature dependent, especially at very low temperature, as illustrated in **Figures 11** and **12**. As the device temperature decreases from 300 K down to 4 K, R_{TH}^* is multiplied by 3 to 6. In FDSOI devices, the BOX tends to confine the heat in the channel, and therefore the total thermal resistance depends on both the thermal conductivity of Si and SiO2, which have different temperature dependence and magnitude (**Figure 13**). R_{TH^*} follows the temperature dependence of the inverse of the silicon dioxide thermal conductivity in the whole range of explored temperatures [57].

Besides considerations over the dominant thermal path in the device, the R_{TH}^* vs. T_{device} plot can be used into thermal model in order to reconstruct the channel

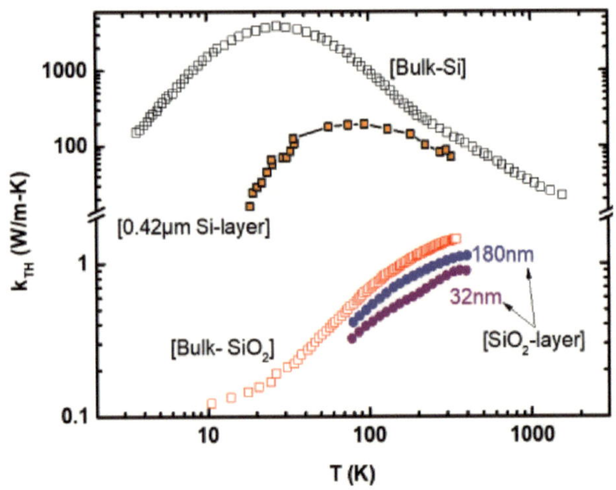

Figure 13.
Thermal conductivity data versus temperature for bulk and Si-layer compared to that for bulk and SiO_2-layer. After Triantopoulos et al. *[55].*

temperature increase ΔT as a function of operating ambient temperature T_{amb} and input power P using the following expression,

$$P = \int_0^{\Delta T} \frac{d\Delta T'}{R_{TH}^{*}(T_{amb} + \Delta T')} \tag{4}$$

Substituting a given analytical expression of $R_{TH}^{*}(T_{device})$ in Eq. (4) the value of ΔT at each T_{amb} and for each value of dissipated power can be calculated (**Figure 14**). This leads in particular to a nonlinear temperature increase of the device with the dissipated power. In this specific low temperature environment, the device temperature can significantly increase and thus highly deviate from the ambient temperature, depending on the applied gate and drain voltages, as illustrated in **Figures 15** and **16**.

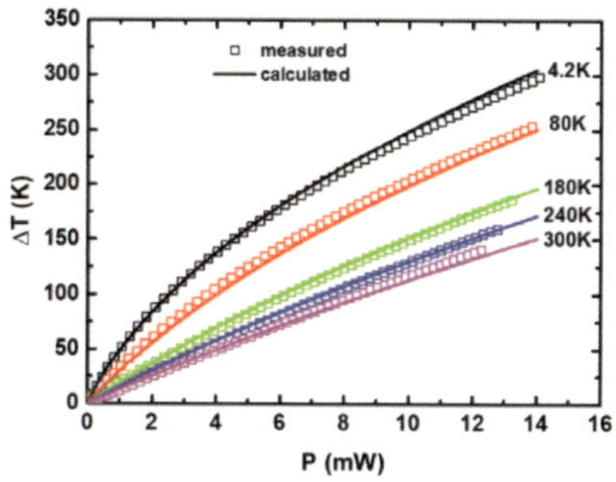

Figure 14.
Calculated channel temperature increase ΔT (line) as a function of the dissipated power P using Eq. (4) and a fitting expression for $R_{TH}^{*}(T_{dev})$. Experimental data (symbols) are also shown for a direct comparison. After Triantopoulos et al. *[55].*

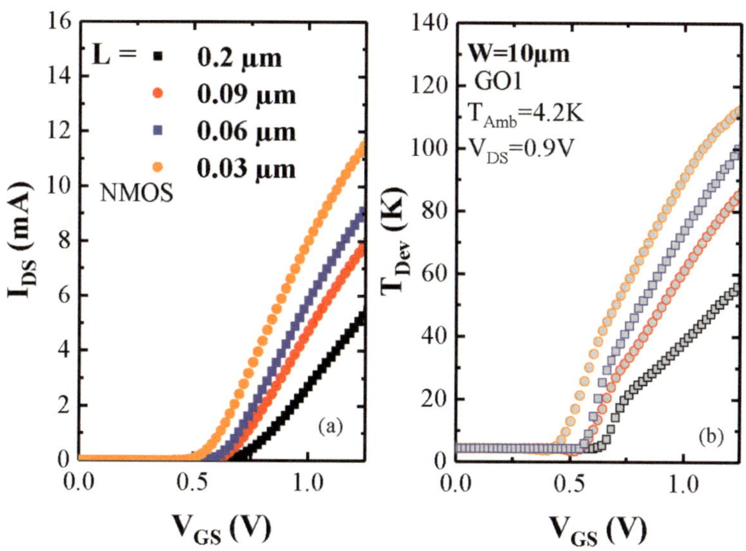

Figure 15.
(a) I_{DS} vs. V_{GS} measured on NMOS at T_{amb} = 4.2 K and V_{DS} = 0.9 V for different gate lengths, and (b) corresponding device temperature, T_{dev} vs. V_{GS}.

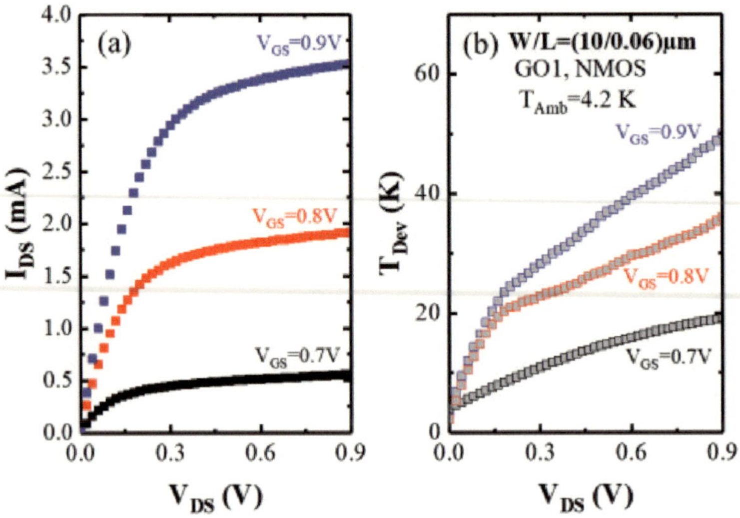

Figure 16.
(a) I_{DS} vs. V_{DS} measured at T_{amb} = 4.2 K on NMOS with L = 60 nm for different V_{GS} values, and (b) corresponding T_{dev}. Vs. V_{DS}.

4. Mismatch and variability properties

The device mismatch is a key property to be known for the development of transistor compact models and the design of electronic circuits [58–60]. This section presents variability results obtained on FDSOI MOSFETs down to 4.2 K. To this end, an integrated on-chip matrix of individually addressable transistors has been used to increase the sample size statistics.

4.1 Devices under test

The measurements were performed on both N- and P-type transistors fabricated using the same 28 nm FDSOI technology as those described in **Section 2.1**. In order to provide statistical analysis on variability and mismatch at low temperature, matrices of transistors were produced with integrated addressability in an approach similar to [61]. An automated measurement system was implemented thanks to the on-chip multiplexed arrangement. The device chips were wire-bonded on a chip carrier connected to a printed circuit board (PCB) and mounted on a dipstick to reach 4.2 K in a liquid helium bath. Each die comprises 512 matched pairs of MOSFETs (256 pairs of RVT plus 256 pairs of LVT) addressable through 10-bits selection (2^{10} = 1024 transistors).

4.2 Threshold voltage variability

Figure 17 shows typical drain current $I_d(V_g)$ characteristics for short channel N-type MOS transistors, at 300 K and at 4.2 K. Twenty four devices were measured for each MOS type at low drain voltage ($|V_d|$ = 50 mV) to illustrate device variability. In **Figure 18** the logarithmic scaled $I_d(V_g)$ emphasizes the subthreshold oscillation variability at low and high drain voltage (V_d = 50 mV and 0.9 V), at 4.2 K. The oscillations observed in the subthreshold current are a known signature of short channel MOSFETs operating at deep cryogenic temperatures, and could result from the presence of impurities in the channel [63, 64]. The threshold voltage was

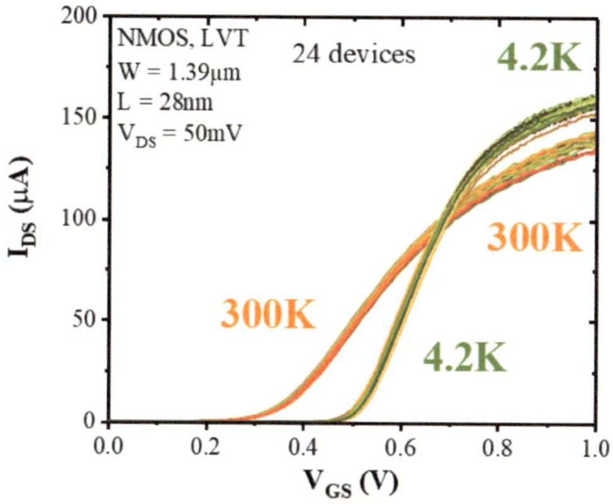

Figure 17.
$I_d(V_g)$ curves for 24 short channel (L = 28 nm) N-type LVT MOSFETs at 4.2 K and 300 K, at V_d = 50 mV. After Cardoso et al. [62].

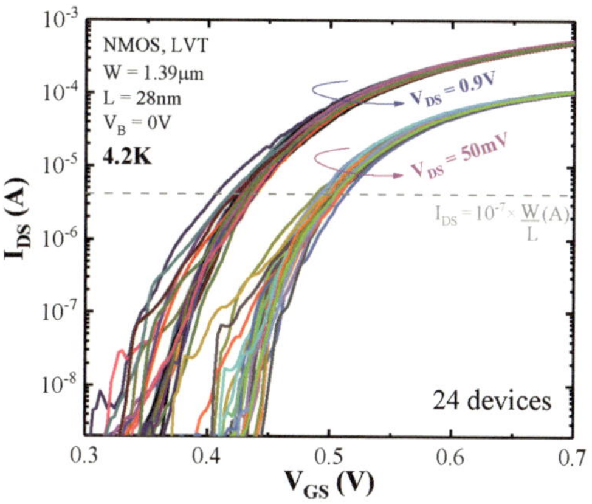

Figure 18.
$I_d(V_g)$ curves for 24 short channel (L = 28 nm) N-type LVT MOSFETs at 4.2 K, at V_d = 50 mV and 0.9 V. After Cardoso et al. [62].

extracted following the constant current criterion, at $I_d = 10^{-7}$ W/L (A). Such current level represents the standard value used for V_{th} extraction, and it is well above the region where the oscillations are mainly identified, as highlighted in **Figure 18** by a dashed line.

Figure 19 shows the Pelgrom plots of the standard deviation of ΔV_{th}, $\sigma_{\Delta VT}$, for NMOS devices (similar results have been obtained for PMOS). It can be seen that $\sigma_{\Delta VT}$ well follows the area scaling linear dependence with respect to $1/\sqrt{W.L}$ at 4.2 K, as it is the case at 300 K, for all channel dimensions explored in this study (1 μm ≤ L ≤ 28 nm, 80 nm ≤ W ≤ 25 μm). This result does not reveal any specific variation with channel width and channel length due to e.g. line edge roughness (LER) for such geometries. From 300 K to 4.2 K, the extracted linear slopes, $\Delta\sigma_{\Delta VT}/\Delta(1/\sqrt{W.L})$, indicates that the threshold voltage mismatch performance degrades by ≈25% for NMOS and PMOS, at $|V_d|$ = 50 mV, when temperature is decreased from 300 K down to 4.2 K. Since the metal gate granularity and the local charges in the gate dielectric are the main sources of threshold voltage variability in FDSOI technology [65], the slight increase of $\sigma_{\Delta VT}$ at 4.2 K may likely be attributed to the increase of interface charge density [63]. Moreover, in **Figure 19**, it can be seen that short channel MOSFETs (L = 28 nm) exhibit higher threshold voltage variability at high drain bias (V_d = 0.9 V), which could be due to Drain Induced Barrier Lowering (DIBL).

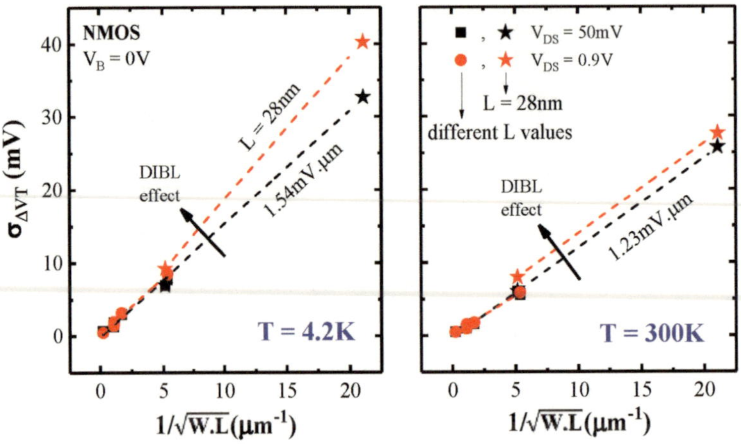

Figure 19.
Pelgrom plot of threshold voltage variability $\sigma_{\Delta VT}$ for NMOS at V_d = 50 mV and 0.9 V, 4.2 K (left) and 300 K (right). After Cardoso et al. [62].

Figure 20 shows the threshold voltage individual mismatch parameter, $A_{\Delta VT} = \sigma_{\Delta VT} \cdot \sqrt{W.L}$, plotted as a function of $1/\sqrt{W.L}$, for 28 nm FDSOI transistors studied in this work and 40 nm bulk MOSFETs from [61], at 300 K and 4.2 K. Despite $A_{\Delta VT}$ degradation at low temperature, FDSOI remains highly competitive compared to bulk technology, mainly due to the suppression of random dopant fluctuation (RDF) induced variability in FDSOI. In **Figure 20**, it can also be observed that $A_{\Delta VT}$ does not exhibit higher values for the short channel MOSFETs (i.e. high $1/\sqrt{W.L}$ values), for which subthreshold oscillations have been observed at low temperature (**Figure 18**). This means that such oscillations do not have a significant impact on the threshold voltage variability, mainly because they occur below the drain current level where the threshold voltage is extracted, as discussed before.

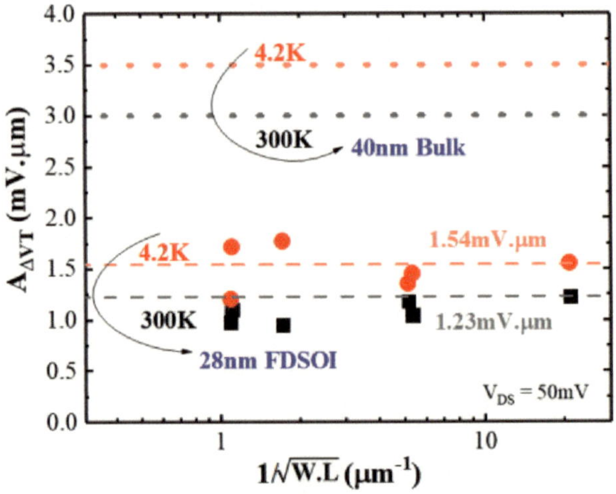

Figure 20.
$A_{\Delta VT}$ versus $1/\sqrt{W.L}$ for NMOS, at V_d = 50 mV, 4.2 K and 300 K. dashed lines indicate the extracted linear slope values from the Pelgrom plots. Dotted lines show typical 40 nm bulk CMOS technology data [61]. After Cardoso et al. *[62].*

4.3 Drain current variability

The drain current variability, $\sigma(\Delta I_d/I_d)$, has also been directly measured on the 28 nm FDSOI transistors studied here and their variations with gate voltage overdrive are shown in **Figure 21** for 300 K and 4.2 K. As is usual, $\sigma(\Delta I_d/I_d)$ is maximized below threshold before to decrease in strong inversion, where it might slightly increase again due to the contribution of access resistance R_s variability [66]. Actually, these variations can be very well fitted by the model of Eq. (5) developed for room temperature:

$$\sigma\left(\frac{\Delta I_d}{I_d}\right)^2 = \left(\frac{g_m}{I_d}\right)^2 .\sigma_{\Delta VT}{}^2 + \left(1 - g_d.R_s\right)^2 .\sigma_{\Delta\beta/\beta}{}^2 + g_d^2.\sigma_{\Delta Rs}{}^2 \tag{5}$$

where g_m is the transconductance and g_d is the output conductance. In this model, the drain current variability is controlled by three matching parameters related respectively to the threshold voltage, $\sigma_{\Delta VT}$, the gain factor $\sigma_{\Delta\beta/\beta}$ (β = W/L. $C_{ox}.\mu_0$, with μ_0 being the low-field carrier mobility) and to the access resistance $\sigma_{\Delta Rs}$. Typical matching parameters extracted from the drain current modeling, as well as their respective contributions are summarized in **Figure 22**. It indicates that there is a slight degradation of variability at low temperature and that the matching

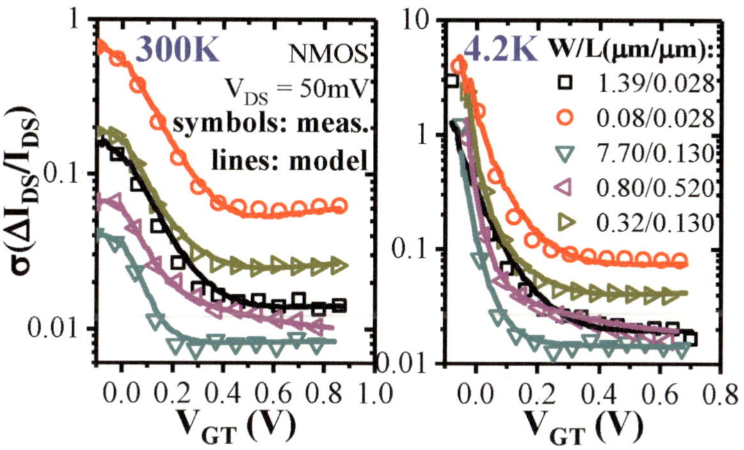

Figure 21.
Measured and modeled $\sigma(\Delta I_d/I_d)$ variations with $V_{gt} = V_g - V_{th}$. $\sigma_{\Delta Rs}$ varies from 0 to 8% of R_s = 377 Ω.µm (T = 300 K) and 266 Ω.µm (T = 4.2 K). After Cardoso et al. *[45].*

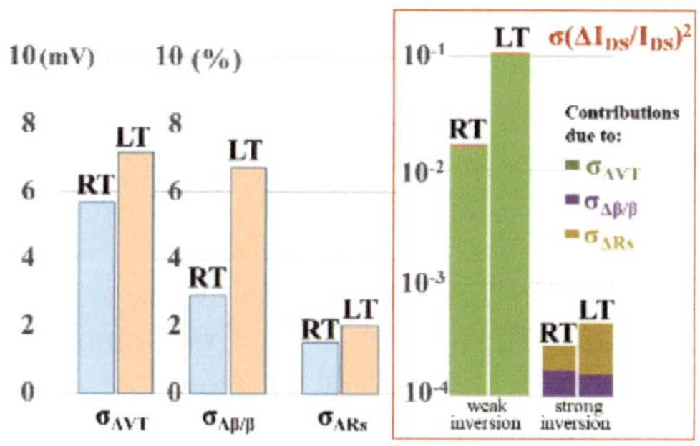

Figure 22.
Summary of matching performance and respective parameter contributions at 300 K (RT) and 4.2 K (LT) for NMOS (W = 1.39 µm, L = 28 nm). After Cardoso et al. [45].

is mainly dominated by threshold voltage variability in weak inversion and by gain factor and access resistance mismatch at strong inversion.

5. Device compact modeling approach

In previous sections, we focus our efforts on understanding individual device physics and variability at cryogenic temperature. In this section, we present typical Poisson-Schrodinger simulation results for the capacitance and charge control in FDSOI structures operated down to deep-cryogenic temperatures and their application for building up an analytical compact model for charge and drain current in FDSOI MOSFET including back biasing effect.

5.1 Poisson-Schrodinger simulations

Poisson-Schrodinger (PS) simulations were conducted after solving self-consistently the Schrodinger and Poisson equations given below:

$$H(\psi) = E \cdot \psi \tag{6}$$

$$\nabla(\varepsilon_r \nabla V) = -\frac{q \cdot n(x)}{\varepsilon_0} \tag{7}$$

with H the Hamiltonian, E the system energy, ψ the electron wave function, ε_0 and ε_r the vacuum and relative silicon permittivity, n the carrier density as a function of position x in the Si channel depth. The electrical potential V, the subband energies $E_{i,j}$ and wave functions $\psi_{i,j}$ for valley j and level i are numerically calculated in a FDSOI structure for given front and back gate voltages. Then, the electron density is obtained after summing the different valleys and subband contributions as:

$$n(x) = \sum_{j=1}^{2} \sum_{i=1}^{i_{max}} g_j A_{2D,j} k_B T \psi_{i,j}^2(x).F_0\left(\frac{E_f - E_{i,j}}{k_B T}\right) \tag{8}$$

where $k_B T$ is the thermal energy, F_0 is the zero-order Fermi-Dirac integral function, E_f the Fermi level, $E_{i,j}$ the subband energy, g_j the valley degeneracy, and $A_{2D,j}$ the 2D density of states of valley j.

It should be noted that in order to compute the PS equations down to very low temperature (1 K), special truncation caution has been taken to avoid numerical overload in the F_0 Fermi integral function accounting for Fermi-Dirac statistics. PS simulations were also possible at 0 K by replacing the F_0 Fermi-Dirac integral function by a Heaviside function, thus mimicking the fully degenerate metallic statistics.

The 1D FDSOI structure used for PS simulation is depicted in **Figure 23**, showing the band diagram across the stack and typical electron density profile in the channel obtained at T = 4 K for a given bias condition.

Figure 24 demonstrates the variations of the inversion charge Q_i in the Si film as a function of front gate voltage V_g with V_b = +3 V, obtained from PS simulations for various temperatures between 0 K and 60 K. A strong increase of the subthreshold slope with temperature dropping can be noticed, reaching infinity at T = 0 K, which is an interesting feature for transistors operating at such low temperatures.

Figure 25 shows the inversion charge control by field effect through the variations of the gate-to-channel capacitance $C_{gc}(V_g) = dQ_i/dV_g$ with front gate voltage

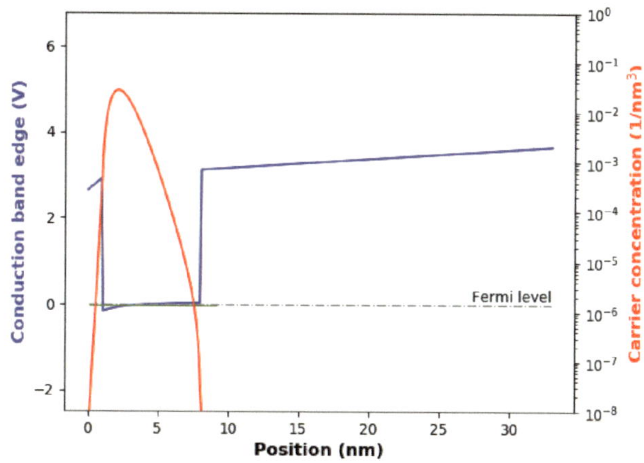

Figure 23.
Typical band diagram and electron distribution from PS simulation for a FDSOI structure (V_g = 1 V, t_{ox} = 1 nm, t_{box} = 25 nm, t_{si} = 7 nm, V_b = 0 V, T = 4 K). After Aouad et al. [41].

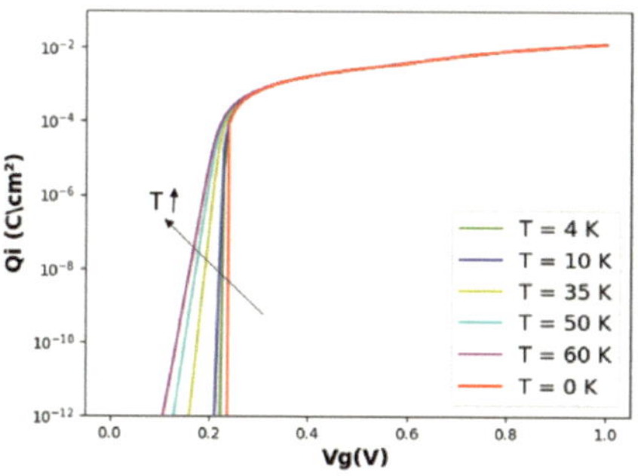

Figure 24.
Inversion charge $Q_i(V_g)$ calculated for different temperatures (t_{ox} = 1 nm, t_{box} = 25 nm, t_{si} = 7 nm, V_b = +3 V). After Aouad et al. [41].

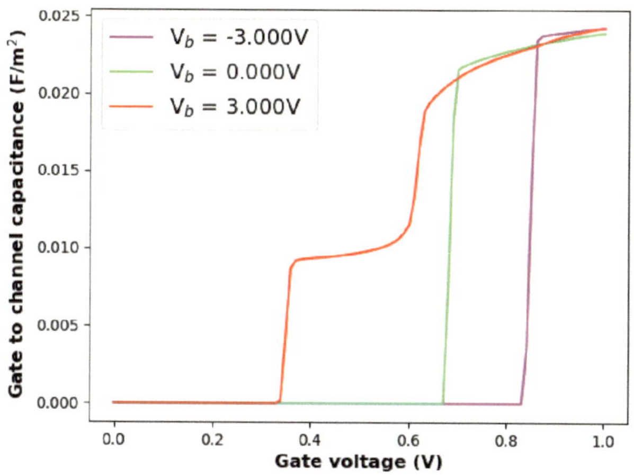

Figure 25.
$C_{gc}(V_g)$ curves for different back biases V_b (t_{ox} = 1 nm, t_{box} = 25 nm, t_{si} = 10 nm, T = 4 K). After Aouad et al. [41].

for various back gate biases V_b. The onset of the back inversion channel for V_b = +3 V is evidenced by an additional plateau in the $C_{gc}(V_g)$ curve, followed by the front channel opening. This effect clearly demonstrates the capacitive coupling, through the silicon channel, between the front gate and the back channel inversion layer, which leads to a lower capacitance.

The impact of temperature on the $C_{gc}(V_g)$ characteristics is shown in **Figure 26**, clearly revealing the rounding of the curves with temperature rise above T = 10 K.

5.2 Compact modeling

Following the PS simulation results, an analytical model has been established considering that front and back channel charges can be evaluated separately at each interface within a single subband approximation with energy level of a triangular potential well [41]. The coupling between the front and back channels is realized owing to the silicon channel capacitance C_{si} and the charge sheet approximation with Fermi-Dirac statistics.

In this case, the charge conservation equations at front and back interfaces are expressed by:

$$V_g = V_{fb} + V_{s1} + \frac{qN_{inv1}}{C_{ox}} + \frac{C_{si}.(V_{s1} - \Delta V(F_1) - V_{s2} + \Delta V(F_2))}{C_{ox}} \quad (9)$$

$$V_b = V_{fb} + V_{s2} + \frac{qN_{inv2}}{C_{box}} + \frac{C_{si}.(V_{s2} - \Delta V(F_2) - V_{s1} + \Delta V(F_1))}{C_{box}} \quad (10)$$

where the front and back interface 2D charge densities read,

$$N_{inv1,2} = A_{2d}.k_B.T.F_0\left[\frac{V_{s1,2} - V_0 - \Delta V(F_{1,2})}{k_B T}\right] \quad (11)$$

where V_{s1} (V_{s2}) is the front (back) interface surface potential, C_{ox} (C_{box}) the front (back) oxide capacitance, C_{si} the silicon film capacitance. The front and back electric field are given by:

$$F_1 = (V_g - V_{s1} - V_{fb})/3t_{ox} \quad (12)$$

$$F_2 = (V_b - V_{s2} - V_{fb})/3t_{box} \quad (13)$$

with the Airy subband potential shift $\Delta V(F) = K \cdot |F + F_0|^{2/3}$ with K = $1.75 \times 10^{-5}\,V^{1/3}\,cm^{2/3}$ [67]. As the film quantization effect is dominating when

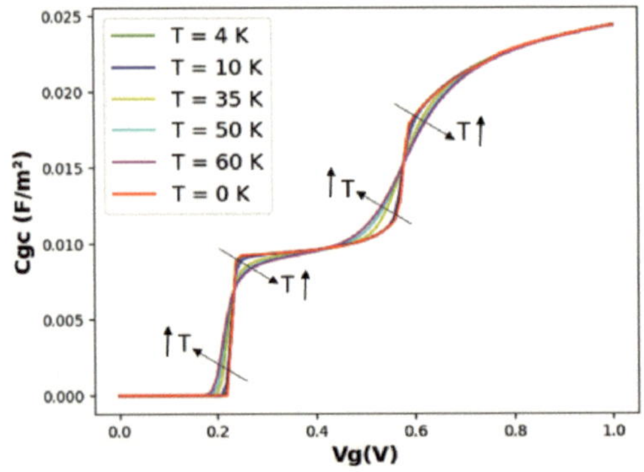

Figure 26.
$C_{gc}(V_g)$ curves for different temperatures (t_{ox} = 1 nm, t_{box} = 25 nm, t_{si} = 10 nm, V_b = +3 V). After Aouad et al. [41].

the electrical field approaches zero, an offset field F_0 is added to the electric field to account for the flat band quantum confinement [41].

Typical $Q_i(V_g)$ and $C_{gc}(V_g)$ characteristics obtained by this Airy-based analytical model are presented in **Figures 27** and **28**, along with the PS simulation results. As can be seen, the compact model provides a good agreement with PS data, emphasizing its physical consistency in terms of charge and capacitance.

The total drain current in the channel can then computed, within the gradual channel approximation, by integrating the channel conductance between source and drain for the front and back channel and by adding their contribution as:

$$I_d = \frac{W}{L}\int_0^{V_d} \mu_{eff1}(Q_{i1}).Q_{i1}(V_{s1} - U_c).dU_c + \frac{W}{L}\int_0^{V_d} \mu_{eff2}(Q_{i2}).Q_{i2}(V_{s2} - U_c).dU_c \tag{14}$$

where U_c is the quasi Fermi level shift between source and drain common to both channels, $Q_{i1,2} = q.N_{inv1,2}$ are the front and back inversion charges obtained from Eq. (11) and $\mu_{eff1,2}$ are the front and back channel effective mobility evaluated separately using Eq. (3). In absence of inter-subband scattering, the drain current calculated using Eqs. (3) and (14) does not exhibit a decrease for V_b = 4 V when the front channel is opening, in contrast to the experimental results discussed in **Section 2.3** (see **Figure 29**). Inter-subband scattering can be taken into account

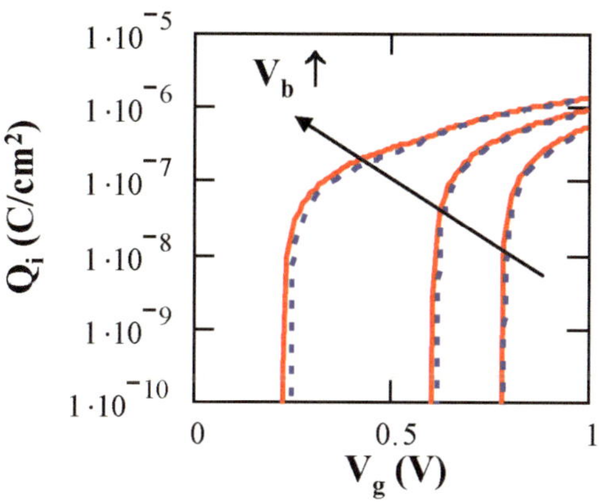

Figure 27.
$Q_i(V_g)$ curves obtained from PS simulations (solid lines) and analytical modeling (dashed lines) for various V_b = −3, 0, +3 V (T = 4 K, t_{ox} = 1 nm, t_{box} = 25 nm, t_{si} = 10 nm). After Aouad et al. [41].

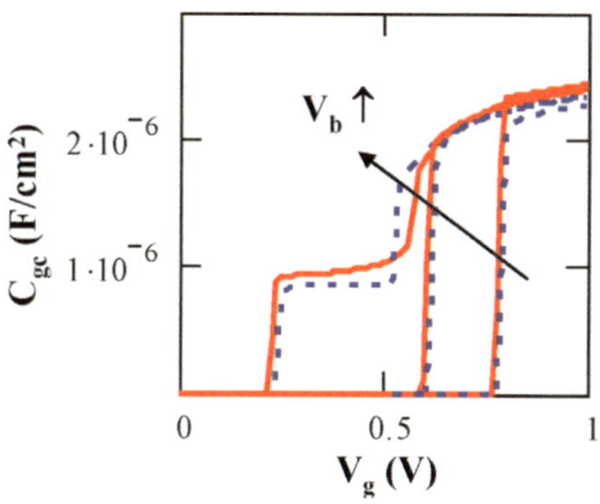

Figure 28.
$C_{gc}(V_g)$ curves obtained from PS simulations (solid lines) and analytical modeling (dashed lines) for various parameters V_b = −3, 0, +3 V (T = 4 K, t_{ox} = 1 nm, t_{box} = 25 nm, t_{si} = 10 nm). After Aouad et al. [41].

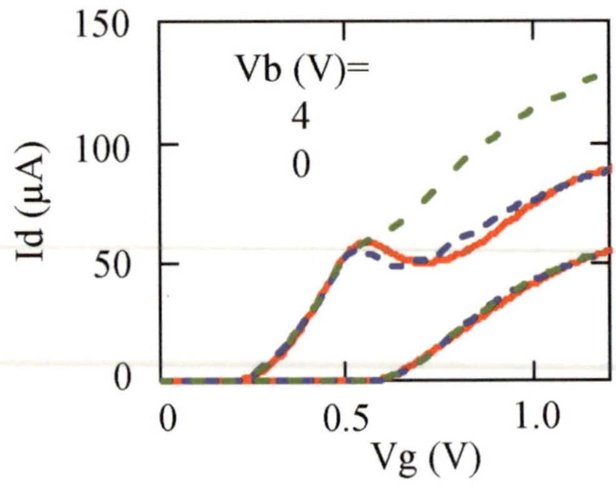

Figure 29.
Drain current vs. front gate voltage V_g: Experimental (red solid line) and modeled with IS (dashed blue line) and modeled without IS (green dashed line) for V_b = 4 V and 0 V at T = 4.2 K.

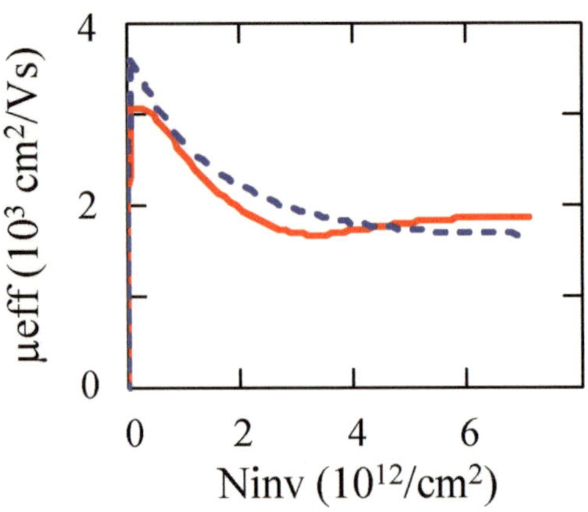

Figure 30.
Experimental (red solid line) and modeled (dashed blue line) back channel mobility μ_{eff} vs. front channel inversion charge density N_{inv1} for V_b = 4 V at T = 4.2 K. model parameters: A = 0.45, b = 0.55 and c = $1.5 \times 10^{12}/cm^2$.

through an additional explicit dependence of the back channel mobility with the front inversion charge density of the form, $\mu_{eff2,IS} = \mu_{eff2} \cdot [a + b.\exp(-N_{inv1}/c)]$, with *a*, *b* and *c* being fitting parameters (**Figure 30**). By this way, the drain current can reasonably be well modeled as shown in **Figure 29** (dashed blue line), inferring the crucial role of remote inter-subband scattering in the back channel mobility.

6. Basic circuit operation at cryogenic temperatures

Although operational cryo-CMOS circuits have been demonstrated down to 30 mK [17, 30, 68–70], unfortunately no mature models are yet available to accurately predict the behavior of passive and active devices at cryogenic temperatures [71, 72]. Due to this lack of compact models at cryogenic temperatures, designers are faced to a blind-design procedure, which reduces the optimization of cryogenic integrated circuits [12, 30, 32, 58, 73–75]. Using the extensive electrical characterizations of single FDSOI transistors at cryogenic temperatures, it is however possible to already design efficient circuits.

Among them oscillators are essential building blocks in many digital and analog circuits. They are required for example to generate a clock signal in the control circuit of quantum computers [30, 76], and so must be also efficient at cryogenic temperature. Here we have electrically characterized ring oscillator (RO) fabricated from 28 nm-FDSOI technology [30, 77]. **Figure 31a** shows the delay per stage of a 101-stages RO as a function of temperature from 300 K down to 4.2 K. Without any back-biases applied on the MOSFETs composing the inverter stages, decreasing the temperature results in slowing down the RO. This can be explained by the threshold voltage shift at cryogenic temperature, which leads to a decrease of the effective current evaluated from the single characteristics of NMOS and PMOS transistors.

The effective drive current I_{EFF}, which is a measure of the current drive of the MOSFET during switching and correlates well to circuit delay, can be defined for a single inverter as [78],

$$I_{EFF} = \left(\frac{1}{I_{EFF,NMOS}} + \frac{1}{I_{EFF,PMOS}} \right)^{-1} \tag{15}$$

with

$$I_{EFF,NMOS/PMOS} = \frac{(I_H + I_L)}{2} \tag{16}$$

where

$$\begin{aligned} I_H &= I_{DS}(V_{GS} = V_{DD}, V_D = V_{DD}/2\,) \\ I_L &= I_{DS}(V_{GS} = V_{DD}/2, V_D = V_{DD}) \end{aligned} \tag{17}$$

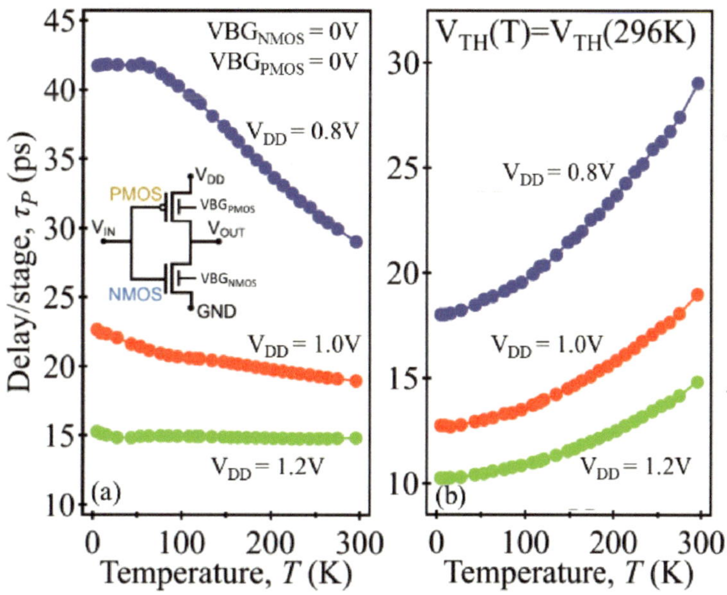

Figure 31.
(a) Delay per stage versus temperature of a 101-stages RO (L = 34 nm, W_{NMOS} = 420 nm, W_{PMOS} = 600 nm) for different supply voltages V_{DD} = 0.8, 1, and 1.2 V showing the RO slowing down due to the increase of V_{TH} at low temperature. (b) Delay per stage versus temperature for V_{DD} = 0.8, 1, and 1.2 V in the case of compensated V_{TH}. The RO speeds up at low temperature due to the carrier mobility enhancement (from Bohuslavskyi et al. [77]).

Figure 32a shows the evolution of I_{EFF} as a function of temperature, in the case where no V_{BG} is applied. We observed that I_{EFF} decreases with temperature, and this decrease is stronger as V_{DD} is decreased (3 decades degradation from 300 K to 4.2 K for V_{DD} = 0.8 V). This I_{EFF} variation is linked with the temperature dependence of I_{DS}-V_{GS} curves. A zero-temperature coefficient point (ZTC), corresponding to a gate voltage for which the drain current exhibits no temperature dependence, is systematically observed on the measured I_{DS} vs. V_{GS} curves, as illustrated in **Figure 33** and already evidenced in **Figures 4** and **17** [79]. For $|V_{GS}| < |V_{ZTC}|$ the drain current decreases as T decreases ($\partial I_{DS}/\partial T|_{VGS\,=\,cte} > 0$), whereas for

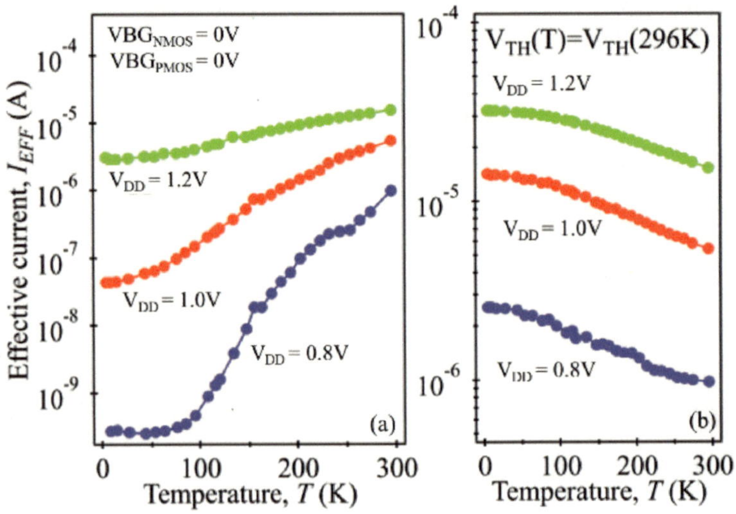

Figure 32.
(a) Effective current I_{EFF} measured on single NMOS and PMOS transistors (L = 34 nm, W_{NMOS} = 210 nm, W_{PMOS} = 300 nm) for different supply voltages V_{DD} = 0.8, 1, and 1.2 V; the effective current decreases as the temperature is reduced. (b) I_{EFF} versus temperature for V_{DD} = 0.8, 1, and 1.2 V in the case of compensated V_{TH}; in that case the effective current increases as the temperature is reduced (from Bohuslavskyi et al. [77]).

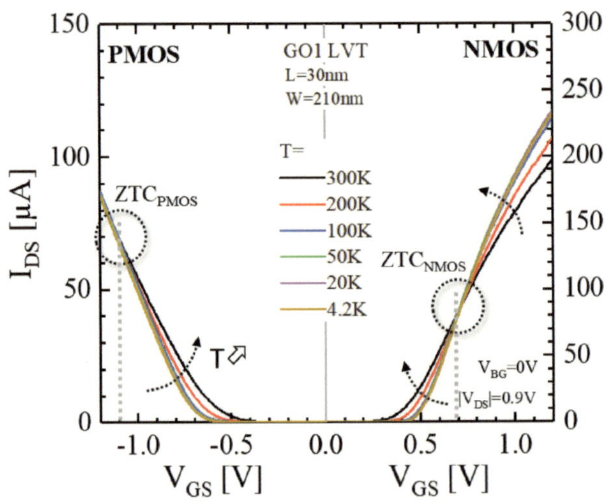

Figure 33.
(a) Drain current I_{DS} measured on single NMOS and PMOS transistors (L = 30 nm, W_{NMOS}=W_{PMOS} = 210 nm) as a function of gate voltage V_{GS} for different temperature from 300 K down to 4.2 K. a zero temperature coefficient (ZTC) point for which the drain current (I_{DS}) is independent of the temperature is evidenced for NMOS and PMOS.

$|V_{GS}| > |V_{ZTC}|$ the drain current exhibits an opposite temperature behavior ($\partial I_{DS}/\partial T|_{VGS\,=\,cte} < 0$).

It is worth noticing that for the pMOS the ZTC point is located at higher $|V_{GS}|$ ($\approx$1.1 V) compared to the nMOS devices ($\approx$0.7 V). The I_{EFF} temperature dependence is mainly driven by the region with positive T-dependence $\partial\, I_{DS}/\partial T|_{VGS\,=\,cte}$, *i.e.* for $|V_{GS}|$ below $|V_{ZTC}|$.

If a back bias V_{BG} is applied, it is possible to shift the threshold voltage back to its room temperature value (**Figure 6**). In that configuration, the drain current I_{DS} increases with the temperature decrease whatever V_{GS} and V_{DS} values, due to mobility and saturation velocity improvement with T decrease at a given $|V_{GS}\text{-}V_{TH}|$ overdrive gate voltage (see **Section 2.4**). Consequently, the effective current I_{EFF} follows the same trend with respect to T (**Figure 32b**). Thus a correctly chosen forward back bias on NMOS and PMOS will lead to a speed-up of the RO as T decreases (**Figure 31b**). At a given temperature, the back biasing V_{BG} allows to tune the frequency as illustrated in **Figure 34**. Finally by playing with the supply voltage V_{DD} and V_{BG} it is also possible to manage power consumption and performance [30, 77]. It has been illustrated at 110mK on a VCO RO (**Figure 35**) where back bias allows switching from low power mode (*e.g.* 27 μW at 2GHz) to high performance mode (*e.g.* 6.9GHz for 268 μW).

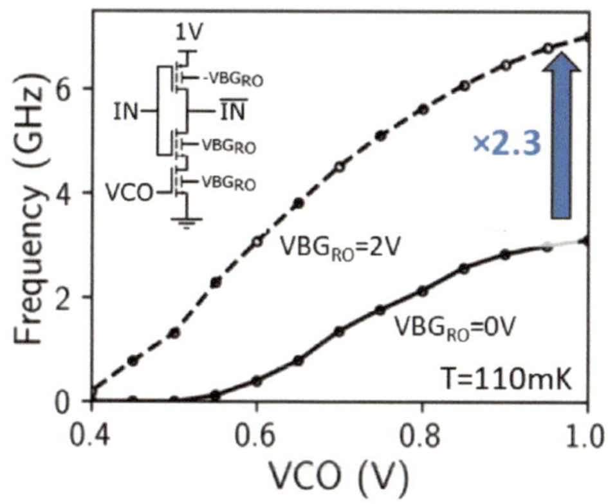

Figure 34.
Oscillating frequency as a function of VCO voltage for a VCO RO (L = 28 nm). Forward Back-biasing increases maximal frequency (from Guevel et al. [30]).

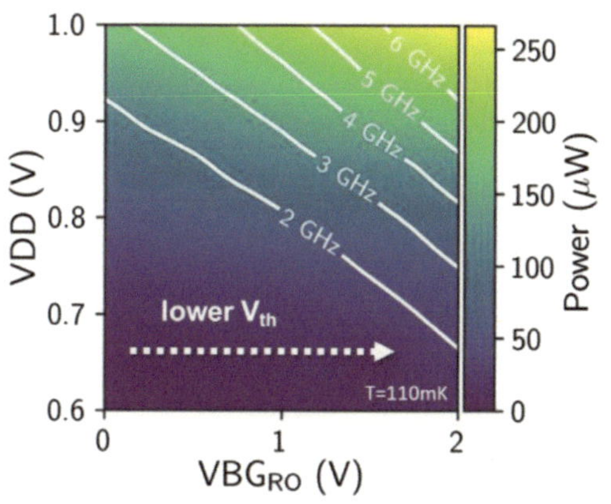

Figure 35.
Power as a function of supply voltage V_{DD} and back bias voltage VBG_{RO} (L = 28 nm). Forward back-biasing decreases power for same frequency (from Guevel et al. [30]).

7. Summary and conclusions

A review of recent results obtained on 28 nm FDSOI transistors operated down to deep cryogenic temperatures has been presented. First, the main device electrical properties in terms of gate capacitance and charge control and drain current transfer characteristics have been discussed along with the temperature dependence of the major MOSFET parameters (threshold voltage, subthreshold swing and mobility). Then, the self-heating phenomena were characterized in details, providing valuable information about the actual device temperature versus power dissipation, as well as the thermal resistance that limits the heat dissipation in the FDSOI structure, especially at low temperature. The matching properties have then been studied owing to threshold voltage and drain current statistical variability analysis, revealing that the mismatch in FDSOI transistors only increases of about 30–40% at deep-cryogenic temperatures. Besides, Poisson-Schrodinger simulations have been carried out with success down to zero Kelvin, giving access to valuable information about the gate charge control in FDSOI structures versus temperature, and, providing physical insight to the development of compact model mandatory for FDSOI circuit design at deep cryogenic temperatures. Finally, the operation of elementary circuits such as ring oscillators and voltage controlled oscillators has been demonstrated in terms of inverter delay and clock frequency down to deep-cryogenic temperatures.

This work highlights the powerful advantage of FDSOI over bulk technology, led by the back biasing capability. It offers in particular an efficient way to manage power consumption and performance, thus mitigating thermal effects, which are crucial aspects in cryo-electronics.

Acknowledgements

This work was partially supported by the French Authorities within the frame of NANO2022 project, and by the ERC Synergy QuCube (Grant No. 810504 — QUCUBE — ERC-2018-SyG), and EU H2020 RIA project SEQUENCE (Grant No. 871764).

Author details

Mikaël Cassé[1] and Gérard Ghibaudo[2*]

1 Université Grenoble Alpes, CEA-LETI, Minatec, Grenoble, France

2 Université Grenoble Alpes, IMEP-LAHC, Minatec, Grenoble, France

*Address all correspondence to: ghibaudo@minatec.inpg.fr

References

[1] R. R. Green, "MOSFET operation at 4.2 K," Review of Scientific Instruments, vol. 39, no. 10, pp. 1495–1497, 1968, doi: 10.1063/1.1683144.

[2] C. G. Rogers, "MOST's at cryogenic temperatures," Solid-State Electronics, vol. 11, no. 11, pp. 1079–1091, 1968, doi: 10.1016/0038-1101(68)90130-5.

[3] W. E. Howard and F. F. Fang, "Low temperature effects in Si FETs," *Solid-State Electronics*, vol. 8, no. 1, pp. 82–83, 1965, doi: https://doi.org/10.1016/0038-1101(65)90011-0.

[4] K. v. Klitzing, G. Dorda, and M. Pepper, "New Method for High-Accuracy Determination of the Fine-Structure Constant Based on Quantized Hall Resistance," Phys. Rev. Lett., vol. 45, no. 6, pp. 494–497, Aug. 1980, doi: 10.1103/PhysRevLett.45.494.

[5] R. K. Kirschman, "Cold electronics: an overview," *Cryogenics*, vol. 25, no. August 1984, pp. 115–122, 1985, doi: 10.1016/0011-2275(85)90036-0.

[6] E. A. Gutiérrez, J. Deen, and C. Claeys, *Low temperature electronics: physics, devices, circuits, and applications*. Academic Press, 2000.

[7] F. Balestra and G. Ghibaudo, *Device and Circuit Cryogenic Operation for Low Temperature Electronics*. 2001.

[8] J. M. Hornibrook *et al.*, "Cryogenic Control Architecture for Large-Scale Quantum Computing," vol. 024010, pp. 1–9, 2015, doi: 10.1103/PhysRevApplied.3.024010.

[9] E. Charbon *et al.*, "Cryo-CMOS for Quantum Computing," *IEEE International Electron Devices Meeting (IEDM)*, pp. 343–346, 2016, doi: 10.1109/ICRC.2017.8123682.

[10] X. Xue *et al.*, "CMOS-based cryogenic control of silicon quantum circuits," 2020, [Online]. Available: http://arxiv.org/abs/2009.14185.

[11] J. P. G. van Dijk *et al.*, "Impact of Classical Control Electronics on Qubit Fidelity," *Physical Review Applied*, vol. 12, no. 4, p. 1, 2019, doi: 10.1103/PhysRevApplied.12.044054.

[12] J. C. Bardin *et al.*, "Design and Characterization of a 28-nm Bulk-CMOS Cryogenic Quantum Controller Dissipating Less Than 2 mW at 3 K," IEEE Journal of Solid-State Circuits, vol. 54, no. 11, pp. 3043–3060, 2019, doi: 10.1109/JSSC.2019.2937234.

[13] E. Charbon, "Cryo-CMOS Electronics for Quantum Computing Applications," in *Proceedings of European Solid-State Device Research Conference*, 2019, pp. 1–6.

[14] D. J. Reilly, "Challenges in Scaling-up the Control Interface of a Quantum Computer," IEDM Technical Digest. IEEE International Electron Devices Meeting, pp. 745–750, 2019.

[15] R. M. Incandela, L. Song, and H. A. R. Homulle, "Characterization and Compact Modeling of Nanometer CMOS Transistors at Deep-Cryogenic Temperatures," *IEEE Journal of the Electron Devices Society*, vol. 6, no. January, pp. 996–1006, 2018, doi: 10.1109/JEDS.2018.2821763.

[16] S. J. Pauka *et al.*, *A Cryogenic Interface for Controlling Many Qubits*. arXiv:1912.01299v1, 2019.

[17] B. Patra *et al.*, "A Scalable Cryo-CMOS 2-to-20GHz Digitally Intensive Controller for 4 × 32 Frequency Multiplexed Spin Qubits/Transmons in 22nm FinFET Technology for Quantum Computers: Visuals," 2020.

[18] F. Sebastiano *et al.*, "Cryo-CMOS Interfaces for Large-Scale Quantum Computers," in *IEEE International Electron Devices Meeting (IEDM)*, 2020, pp. 525–528.

[19] F. Balestra and G. Ghibaudo, "Physics and performance of nanoscale semiconductor devices at cryogenic temperatures," *Semiconductor Science and Technology*, vol. 32, no. 2, 2017, doi: 10.1088/1361-6641/32/2/023002.

[20] B. Dierickx, L. Warmerdam, E. R. Simoen, J. Vermeiren, and C. Claeys, "Model for hysteresis and kink behavior of MOS transistors operating at 4.2 K," IEEE Transactions on Electron Devices, vol. 35, no. 7, pp. 1120–1125, 1988, doi: 10.1109/16.3372.

[21] E. Simoen, B. Dierickx, L. Warmerdam, J. Vermeiren, and C. Claeys, "Freeze-Out Effects on NMOS Transistor Characteristics at 4.2 K," IEEE Transactions on Electron Devices, vol. 36, no. 6, pp. 1155–1161, 1989, doi: 10.1109/16.24362.

[22] F. Balestra and G. Ghibaudo, "Brief review of the MOS device physics for low temperature electronics," Solid-State Electronics, vol. 37, no. 12, pp. 1967–1975, 1994.

[23] T. Lu, Z. Li, C. Luo, J. Xu, W. Kong, and G. Guo, "Characterization and Modeling of 0.18μm Bulk CMOS Technology at sub-Kelvin Temperature," *IEEE Journal of the Electron Devices Society*, vol. 8, no. June, pp. 1–1, 2020, doi: 10.1109/jeds.2020.3015265.

[24] A. Rhouni, O. Gevin, X. de la Broïse, J. L. Sauvageot, V. Revéret, and L. Rodriguez, "First-ever test and characterization of the AMS standard bulk 0.35 μm CMOS technology at sub-kelvin temperatures," Journal of Physics: Conference Series, vol. 834, p. 012005, May 2017, doi: 10.1088/1742-6596/834/1/012005.

[25] R. M. Incandela, L. Song, H. A. R. Homulle, F. Sebastiano, E. Charbon, and A. Vladimirescu, "Nanometer CMOS Characterization and Compact Modeling at Deep-Cryogenic Temperatures," no. 10, pp. 58–61, 2017.

[26] A. Beckers, F. Jazaeri, A. Ruffino, C. Bruschini, A. Baschirotto, and C. Enz, "Cryogenic characterization of 28 nm bulk CMOS technology for quantum computing," in *Proceedings of European Solid-State Device Research Conference*, 2017, pp. 62–65, doi: 10.1109/ESSDERC.2017.8066592.

[27] B. Cretu, D. Boudier, E. Simoen, A. Veloso, and N. Collaert, "Assessment of DC and low-frequency noise performances of triple-gate FinFETs at cryogenic temperatures," *Semiconductor Science and Technology*, vol. 31, no. 12, 2016, doi: 10.1088/0268-1242/31/12/124006.

[28] M. Shin *et al.*, "Low temperature characterization of 14nm FDSOI CMOS devices," in *International Workshop on Low Temperature Electronics (WOLTE)*, Grenoble, France, 2014, pp. 29–32, doi: 10.1109/WOLTE.2014.6881018.

[29] H. Bohuslavskyi *et al.*, "28nm Fully-depleted SOI technology: Cryogenic control electronics for quantum computing," in *IEEE Silicon Nanoelectronics Workshop (SNW)*, 2017, pp. 143–144, doi: 10.23919/SNW.2017.8242338.

[30] L. L. Guevel *et al.*, "A 110mK 295 μW 28nm FDSOI CMOS Quantum Integrated Circuit with a 2.8GHz Excitation and nA Current Sensing of an On-Chip Double Quantum Dot: Visuals," 2020.

[31] L. Nyssens *et al.*, "28-nm FD-SOI CMOS RF Figures of Merit down to 4.2 K," *IEEE Journal of the Electron Devices Society*, vol. 8, no. June, pp. 646–654, 2020, doi: 10.1109/JEDS.2020.3002201.

[32] S. Bonen *et al.*, "Cryogenic Characterization of 22-nm FDSOI CMOS Technology for Quantum Computing ICs," IEEE Electron Device Letters, vol. 40, no. 1, pp. 127–130, 2019, doi: 10.1109/LED.2018.2880303.

[33] H. J. Park, M. Bawedin, H. G. Choi, and S. Cristoloveanu, "Kink effect in ultrathin FDSOI MOSFETs," *Solid-State Electronics*, vol. 143, pp. 33–40, 2018, doi: https://doi.org/10.1016/j.sse.2017.12.002.

[34] C. Enz, A. Beckers, and F. Jazaeri, "Cryo-CMOS Compact Modeling," *IEEE International Electron Devices Meeting (IEDM)*, no. 871764, pp. 529–532, 2020.

[35] J.-P. Noel *et al.*, "Multi-VT UTBB FDSOI Device Architectures for Low-Power CMOS Circuit," IEEE Transactions on Electron Devices, vol. 58, no. 8, pp. 2473–2482, 2011.

[36] M. Kantner and T. Koprucki, "Numerical simulation of carrier transport in semiconductor devices at cryogenic temperatures," *Optical and Quantum Electronics*, vol. 48, no. 12, p. 543, Nov. 2016, doi: 10.1007/s11082-016-0817-2.

[37] S. S. Sesnic and G. R. Craig, "Thermal Effects in JFET and MOSFET at Cryogenic Temperatures," IEEE Transactions on Electron Devices, vol. 19, no. 8, pp. 933–942, 1972.

[38] D. P. Foty and S. L. Titcomb, "Thermal Effects in n-Channel Enhancement MOSFET's Operated at Cryogenic Temperatures," IEEE Transactions on Electron Devices, vol. 34, no. 1, pp. 107–113, 1987, doi: 10.1109/T-ED.1987.22892.

[39] N. Planes *et al.*, "28nm FDSOI technology platform for high-speed low-voltage digital applications," in *Symposium on VLSI Technology,* Digest of Technical Papers, 2012, vol. 33, no. 4, pp. 133–134, doi: 10.1109/VLSIT.2012.6242497.

[40] B. Cardoso Paz *et al.*, "Electrostatics and channel coupling on 28 nm FD-SOI for cryogenic applications," in *Joint International EUROSOI Workshop and International Conference on Ultimate Integration on Silicon (EUROSOI-ULIS)*, 2020, p. 1.

[41] M. Aouad, S. Martinie, T. Poiroux, and G. Ghibaudo, "Poisson-Schrodinger simulation of inversion charge in FDSOI MOSFET down to 0K – Towards compact modeling for cryo CMOS application," in *Joint International EUROSOI Workshop and International Conference on Ultimate Integration on Silicon (EUROSOI-ULIS)*, Caen, France, 2020, p. 1.

[42] T. A. Karatsori, G. Ghibaudo, C. Dimitriadis, and C. Theodorou, "Influence of AC signal oscillator level on effective mobility measurement by split C-V technique in MOSFETs," *Electronics Letters*, vol. 52, 2016, doi: 10.1049/el.2016.1715.

[43] B. Cardoso Paz *et al.*, "Performance and Low-Frequency Noise of 22-nm FDSOI down to 4.2 K for Cryogenic Applications," *IEEE Transactions on Electron Devices*, vol. 67, no. 11, 2020, doi: 10.1109/TED.2020.3021999.

[44] Y. Taur and T. H. Ning, *Fundamentals of Modern VLSI Devices*, vol. 25. 2002.

[45] B. Cardoso Paz *et al.*, "Variability Evaluation of 28nm FD-SOI Technology at Cryogenic Temperatures down to 100mK for Quantum Computing," in *Symposium on VLSI Technology, Digest of Technical Papers*, 2020, p. TN2.1.

[46] E. Arnold, "Disorder-induced carrier localization in silicon surface inversion layers," Applied Physics Letters, vol. 25, no. 12, pp. 705–707, 1974, doi: 10.1063/1.1655369.

[47] H. Bohuslavskyi *et al.*, "Cryogenic Subthreshold Swing Saturation in FD-SOI MOSFETs Described With Band Broadening," IEEE Electron Device Letters, vol. 40, no. 5, pp. 784–787, 2019, doi: 10.1109/LED.2019.2903111.

[48] G. Ghibaudo, M. Aouad, M. Cassé, S. Martinie, T. Poiroux, and F. Balestra, "On the modelling of temperature dependence of subthreshold swing in MOSFETs down to cryogenic temperature," *Solid-State Electronics*, vol. 170, no. April, p. 107820, 2020, doi: 10.1016/j.sse.2020.107820.

[49] B. Cardoso Paz, "Front and back channels coupling and transport on 28 nm FD-SOI MOSFETs down to liquid-He temperature," *Solid State Electronics*, 2021.

[50] M. Cassé *et al.*, "Evidence of 2D intersubband scattering in thin film fully depleted silicon-on-insulator transistors operating at 4.2 K," Applied Physics Letters, vol. 116, p. 243502, 2020, doi: 10.1063/5.0007100.

[51] B. M. Tenbroek, M. S. L. Lee, W. Redman-white, R. J. T. Bunyan, and M. J. Uren, "Self -Heating Effects in SOI MOSFET's and their measurement by small signal conductance techniques," IEEE Transactions on Electron Devices, vol. 43, no. 12, pp. 2240–2248, 1996.

[52] C. Xu, S. K. Kolluri, and K. Endo, "Analytical Thermal Model for Self-Heating in Advanced FinFET Devices With Implications for Design and Reliability," IEEE Trans on Computer-Aided Design Int. Circ. Syst., vol. 32, no. 7, pp. 1045–1058, 2013.

[53] C. Prasad, L. Jiang, D. Singh, M. Ag, J. Thomas, and P. Vandervoorn, "Self-heat Reliabili ity Considerations on Int tel' s 22nm Tri i-Gate Technology," in *IEEE International Reliability Physics Symposium (IRPS)*, 2013, pp. 1–5.

[54] C. Prasad, S. Ramey, and L. Jiang, "Self-heating in advanced CMOS technologies," in *IEEE International Reliability Physics Symposium (IRPS)*, 2017, pp. 6A–4.1–6A–4.7, doi: 10.1109/IRPS.2017.7936336.

[55] K. Triantopoulos *et al.*, "Self-Heating Effect in FDSOI Transistors Down to Cryogenic Operation at 4.2 K," *IEEE Transactions on Electron Devices*, vol. 66, no. 8, 2019, doi: 10.1109/TED.2019.2919924.

[56] P. G. Mautry and J. Trager, "Self-heating and Temperature Measurement in Sub-μm-MOSFETs," *ESSDERC '89: 19th European Solid State Device Research Conference*, pp. 675–678, 1989.

[57] T. Takahashi, T. Matsuki, T. Shinada, Y. Inoue, and K. Uchida, "Direct evaluation of self-heating effects in bulk and ultra-thin BOX SOI MOSFETs using four-terminal gate resistance technique," IEEE Journal of the Electron Devices Society, vol. 4, no. 5, pp. 365–373, 2016, doi: 10.1109/JEDS.2016.2568261.

[58] A. Beckers, F. Jazaeri, and C. Enz, "Cryogenic MOS Transistor Model," IEEE Transactions on Electron Devices, vol. 65, no. 9, pp. 3617–3625, Sep. 2018, doi: 10.1109/TED.2018.2854701.

[59] R. M. Incandela, L. Song, H. Homulle, E. Charbon, A. Vladimirescu, and F. Sebastiano, "Characterization and Compact Modeling of Nanometer CMOS Transistors at Deep-Cryogenic Temperatures," IEEE Journal of the Electron Devices Society, vol. 6, pp. 996–1006, 2018, doi: 10.1109/JEDS.2018.2821763.

[60] A. Beckers, F. Jazaeri, H. Bohuslavskyi, L. Hutin, S. De Franceschi, and C. Enz, "Design-oriented Modeling of 28 nm FDSOI CMOS Technology down to 4.2 K for Quantum Computing," *2018 Joint International EUROSOI Workshop and International Conference on Ultimate Integration on Silicon (EUROSOI-ULIS)*,

pp. 1–4, Mar. 2018, doi: 10.1109/ULIS.2018.8354742.

[61] P. A. 't Hart, J. P. G. van Dijk, M. Babaie, E. Charbon, A. Vladimircscu, and F. Sebastiano, "Characterization and Model Validation of Mismatch in Nanometer CMOS at Cryogenic Temperatures," in *2018 48th European Solid-State Device Research Conference (ESSDERC)*, Sep. 2018, pp. 246–249, doi: 10.1109/ESSDERC.2018.8486859.

[62] B. Cardoso Paz *et al.*, "Integrated Variability Measurements of 28 nm FDSOI MOSFETs down to 4.2 K for Cryogenic CMOS Applications," in *IEEE International Conference on Microelectronic Test Structures*, 2020, vol. 2020-May, pp. 8–12, doi: 10.1109/ICMTS48187.2020.9107906.

[63] H. Bohuslavskyi *et al.*, "Cryogenic Subthreshold Swing Saturation in FD-SOI MOSFETs described with Band Broadening," IEEE Electron Device Letters, pp. 1–1, 2019, doi: 10.1109/LED.2019.2903111.

[64] A. Beckers, F. Jazaeri, and C. Enz, "Revised theoretical limit of the subthreshold swing in field-effect transistors," *arXiv:1811.09146 [cond-mat]*, Nov. 2018, Accessed: Aug. 08, 2019. [Online]. Available: http://arxiv.org/abs/1811.09146.

[65] J. Mazurier, O. Weber, F. Andrieu, C. L. Royer, O. Faynot, and M. Vinet, "Variability of planar Ultra-Thin Body and Buried oxide (UTBB) FDSOI MOSFETs," in *2014 IEEE International Conference on IC Design Technology*, May 2014, pp. 1–4, doi: 10.1109/ICICDT.2014.6838617.

[66] E. G. Ioannidis, C. G. Theodorou, S. Haendler, E. Josse, C. A. Dimitriadis, and G. Ghibaudo, "Impact of Source – Drain Series Resistance on Drain Current Mismatch in Advanced Fully Depleted SOI n-MOSFETs," IEEE Electron Device Letters, vol. 36, no. 5, pp. 433–435, 2015, doi: 10.1109/LED.2015.2411289.

[67] A. P. Gnädinger and H. E. Talley, "Quantum mechanical calculation of the carrier distribution and the thickness of the inversion layer of a MOS field-effect transistor," *Solid-State Electronics*, vol. 13, no. 9, pp. 1301–1309, 1970, doi: https://doi.org/10.1016/0038-1101(70)90027-4.

[68] S. R. Ekanayake, T. Lehmann, A. S. Dzurak, R. G. Clark, and A. Brawley, "Characterization of SOS-CMOS FETs at low temperatures for the design of integrated circuits for quantum bit control and readout," IEEE Transactions on Electron Devices, vol. 57, no. 2, pp. 539–547, 2010, doi: 10.1109/TED.2009.2037381.

[69] S. J. Pauka *et al.*, "Characterizing Quantum Devices at Scale with Custom Cryo-CMOS," *Physical Review Applied*, vol. 13, no. 5, p. 1, 2020, doi: 10.1103/PhysRevApplied.13.054072.

[70] I. D. Conway Lamb *et al.*, "An FPGA-based instrumentation platform for use at deep cryogenic temperatures," *Review of Scientific Instruments*, vol. 87, no. 1, p. 014701, 2016, doi: 10.1063/1.4939094.

[71] A. Beckers, F. Jazaeri, H. Bohuslavskyi, L. Hutin, S. D. Franceschi, and C. Enz, "Characterization and modeling of 28-nm FDSOI CMOS technology down to cryogenic temperatures ☆," Solid State Electronics, vol. 159, no. 688539, pp. 106–115, 2019, doi: 10.1016/j.sse.2019.03.033.

[72] C. Enz, A. Beckers, and F. Jazaeri, "Cryo-CMOS Compact Modeling," in *IEEE International Electron Devices Meeting (IEDM)*, 2020, no. 871764.

[73] E. D. Buchanan, D. J. Benford, J. B. Forgione, S. H. Moseley, and E. J. Wollack, "Cryogenic applications of

commercial electronic components," Cryogenics, vol. 52, no. 10, pp. 550–556, 2012, doi: 10.1016/j.cryogenics.2012.06.017.

[74] E. Schriek, F. Sebastiano, and E. Charbon, "A Cryo-CMOS Digital Cell Library for Quantum Computing Applications," IEEE Solid-State Circuits Letters, vol. 3, pp. 310–313, 2020, doi: 10.1109/LSSC.2020.3017705.

[75] J. P. G. van Dijk *et al.*, "A Scalable Cryo-CMOS Controller for the Wideband Frequency-Multiplexed Control of Spin-Qubits and Transmons," IEEE Journal of Solid-State Circuits, vol. 55, no. 11, pp. 2930–2946, 2020.

[76] J. Gong, Y. Chen, F. Sebastiano, E. Charbon, and M. Babaie, "A 200dB FOM 4-5GHz Cryogenic Oscillator with an Automatic Common-Mode Resonance Calibration for Quantum Computing Applications: Visuals," 2020.

[77] H. Bohuslavskyi *et al.*, "Cryogenic Characterization of 28-nm FD-SOI Ring Oscillators With Energy Efficiency Optimization," IEEE Transactions on Electron Devices, vol. 65, no. 9, pp. 3682–3688, 2018.

[78] M. H. Na, E. J. Nowak, W. Haensch, and J. Cai, "The effective drive current in CMOS inverters," *IEEE International Electron Devices Meeting (IEDM)*, pp. 121–124, 2002, doi: 10.1109/IEDM.2002.1175793.

[79] K. Triantopoulos, M. Cassé, L. Brunet, P. Batude, G. Reimbold, and G. Ghibaudo, "Self-heating assessment and cold current extraction in FDSOI MOSFETs," San Francisco, 2017, doi: 10.1109/S3S.2017.8309239.

Chapter 3

Effect of Roughness Elements on the Evolution of Thermal Stratification in a Cryogenic Propellant Tank

S.B. Vishnu and Biju T. Kuzhiveli

Abstract

The cryogenic propulsion era started with the use of liquid rockets. These rocket engines use propellants in liquid form with reasonably high density, allowing reduced tank size with a high mass ratio. Cryogenic engines are designed for liquid fuels that have to be held in liquid form at cryogenic temperature and gas at normal temperatures. Since propellants are stored at their boiling temperature or subcooled condition, minimal heat infiltration itself causes thermal stratification and self-pressurization. Due to stratification, the state of propellant inside the tank varies, and it is essential to keep the propellant properties in a predefined state for restarting the cryogenic engine after the coast phase. The propellant's condition at the inlet of the propellant feed system or turbo pump must fall within a narrow range. If the inlet temperature is above the cavitation value, cavitation will likely to happen to result in the probable destruction of the flight vehicle. The present work aims to find an effective method to reduce the stratification phenomenon in a cryogenic storage tank. From previous studies, it is observed that the shape of the inner wall surface of the storage tank plays an essential role in the development of the stratified layer. A CFD model is established to predict the rate of self-pressurization in a liquid hydrogen container. The Volume of Fluid (VOF) method is used to predict the liquid–vapor interface movement, and the Lee phase change model is adopted for evaporation and condensation calculations. A detailed study has been conducted on a cylindrical storage tank with an iso grid and rib structure. The development of the stratified layer in the presence of iso grid and ribs are entirely different. The buoyancy-driven free convection flow over iso grid structure result in velocity and temperature profile that differs significantly from a smooth wall case. The thermal boundary layer was always more significant for iso grid type obstruction, and these obstructions induces streamline deflection and recirculation zones, which enhances heat transfer to bulk liquid. A larger self-pressurization rate is observed for tanks with an iso grid structure. The presence of ribs results in the reduction of upward buoyancy flow near the tank surface, whereas streamline deflection and recirculation zones were also perceptible. As the number of ribs increases, it nullifies the effect of the formation of recirculation zones. Finally, a maximum reduction of 32.89% in the self-pressurization rate is achieved with the incorporation of the rib structure in the tank wall.

Keywords: Thermal stratification, self-pressurization, liquid hydrogen, iso grid, cryogenic tank

1. Introduction

Stratification in cryogenic liquid storage systems is a complicated yet inexorable thermodynamic phenomenon involving a combination of heat and mass transfer. Owing to the very low boiling point, they are susceptible to heat ingress from the ambient. Cryogenic propellant tanks undergo typical operation sequences when preparing for launch, including tank filling, chilling, boil-off, level correction, tank pressure and hold until lift-off. The time duration between tank pressurization and lift-off is called ground parking period. During ground parking, the heat ingress is due to the large temperature gradient existing between the storage and atmospheric temperature. An overwhelming increase in heat ingress occurs due to aerodynamic heating (during flight) and space radiation, although not as significant as the former, during coast phase. The insulation provided to the propellant tanks is foam which is relatively less effective compared to vacuum or multi-layered insulation (MLI). This heat leakage raises the temperature of liquid adjacent to the walls inducing natural convection currents. The heated liquid starts flowing up due to buoyancy and accumulates at the liquid-vapor interface creating an axial temperature gradient called thermal stratification. The depth of this stratified layer increases with time.

Consequently, the tank pressure keeps increasing due to vaporization. This demands the proper design of venting devices and insulation system. Hence thermal stratification is a crucial design criterion for designing rocket fuel tanks. A schematic for thermal stratification is illustrated in **Figure 1**.

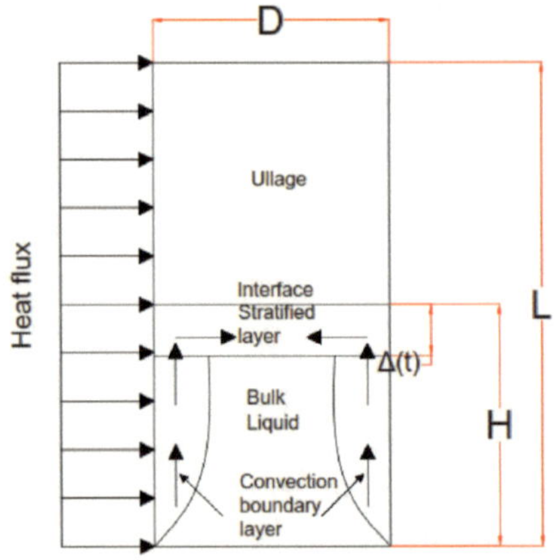

Figure 1.
Schematic of stratification phenomenon in a cylindrical tank.

2. Literature review

A wide variety of experimental and numerical studies have been conducted and reported by researchers across the globe on thermal stratification and self-pressurization of a cryogenic storage vessel. Tatom et al. [1] conducted an experimental investigation on a 500 - gal liquid hydrogen storage tank to provide stratification test data. The main objective of the study was to find out the effect of bottom heating on stratification. It was found that the controlled bottom heating can reduce the degree of stratification by transferring a large fraction of sidewall heat flux to bulk liquid. Schmidt et al. [2] conducted an experimental investigation to study pressurization and stratification of an LH2 tank and compared the results with the theoretical model. It is concluded that as interface temperature increases, the ullage pressure in the tank also increases, which results in more heat transfer to the stratified layer. The amount of additional heat transfer can be calculated using the theoretical model, which helps to optimize the pressure level within the storage tank. Ruder et al. [3] developed a mathematical method to determine the temperature profile inside a cryogenic storage tank under pressurized condition. They developed an empirical relation to represent the temperature profile which is similar in shape to a Gaussian probability distribution.

Several experimental kinds of researches on the evolution of thermal stratification and self-pressurization in a cryogenic storage tank have been reported in the literature. Aydelott et al. [4] developed a non-venting 9-inch diameter spherical container partially filled with liquid hydrogen for self-pressurization tests. The effects of filling per cent, heat flux, top heating, bottom and uniform heating were studied. It was found that the self-pressurization rate in the storage tank was mainly a function of the heating configuration with the per cent filling and heat-transfer rate playing a secondary role.

Ji et al. [5] conducted experiments on a heated container of liquid and developed various dimensionless parameters associated with thermal stratification. By comparing the dimensionless pressure and temperature, the quality of scaling was verified and concluded that parameters like tank pressure, bulk liquid temperature and the surface temperature could be scaled with dimensionless parameters. To study the self pressurization of a spherical liquid hydrogen tank, Hasan et al. [6] conducted an experimental investigation. The results showed that the initial conditions of the storage tank play a considerable role in self-pressurization. If the tank had previously experienced a steady boil-off rate due to long period venting, the self-pressurization rate seems to be lower and rapid pressurization rate was observed for the tank which is not at a steady venting condition. Yamaji et al. [7] conducted an experimental and numerical investigation of thermal stratification and plume mixing. Series of experiments were carried out using PIV and thermocouple measurements, and the developed CFX model reproduced the measured velocity distributions and temperature relatively well.

Gursu et al. [8] developed three different pressure rise models to predict thermal stratification and self-pressurization performance in cryogenic storage tanks; a homogeneous model, a surface evaporation model and a thermal stratification model. The thermal stratification model succeeds in predicting self-pressurization, whereas the other two models which are collectively the isothermal models could not predict the tank pressurization accurately enough. The flow pattern and thermal stratification of a cryogenic cylindrical tank were numerically studied by Chin-Shun Lin et al. [9]. The tank sidewall was subjected to either a uniform heat-flux or two discrete levels of uniform heat-flux at the upper and lower halves of the tank wall. The tank bottom was kept at a constant temperature controlled by the heat exchanger of a thermodynamic vent system. They successfully solved dimensionless steady-state conservation equations by a finite-difference method. Li et al. [10] developed a numerical model to study the thermodynamic effect of heat in-leak into a cryogenic tank and validated the model with experimentation. In the upper part of the liquid, a steady vortex-like region is developed as the heat inleak starts. A large temperature difference in the vertical direction was visible but little temperature difference in radial direction because the radial flow is predominant in that region. It was concluded that thermal stratification exists only in sub-cooled liquid and heat in-leak depends on the void fraction of fluid close to the tank wall. A calculation model is developed by Liu et al. [11] to predict self-pressurization and thermal stratification phenomenon of a liquid hydrogen storage tank. The effect of gravity level, fill level and temperature of the wall on the development of stratification were analyzed using the developed model. The effect of thermal aspect ratio on the self-pressurization is studied experimentally by Kang et al. [12]. A double wall vacuum insulated liquid nitrogen storage tank is developed for experimental investigation and the results were validated with the homogeneous model. Because of the thermal stratification, the experimental results slightly differ from that predicted by homogeneous model. The results indicate that the thermal stratification is highly correlated with the thermal aspect ratio.

The effect of rib shape and material thermal conductivity on the development of stratification was numerically studied by Fu et al. [13]. They considered cylindrical ribbed tank with 50% fill volume and rib shapes of rectangular and semi-circular. Tank pressurization rate was found to be lower for rib materials of low thermal conductivity. Semi-circular ribbed tank underwent lesser self-pressurization in comparison with the rectangular ribbed case for same rib cross-sectional area and locations.

Khurana et al. [14] carried out numerical investigations to minimize the thermal stratification in LH2 tanks. By providing transverse wall ribs on the inner surface of the cylindrical tank, they succeeded in achieving a 30% reduction in the stratification parameter. A delayed stratification, as well as lesser natural heat transfer coefficient, is observed for the tank with the ribbed inner surface than smooth wall tank. Polideri et al. [15] conducted an experimental investigation to study transient natural convection on a vertical ribbed wall. They reported a reduction in convective heat transfer coefficient below the initial rib and enhancement past the last one. A reduction in heat transfer performance was reported by Tanda et al. [16] for a case of natural convective air flow over a heated ribbed plate. To obtain the distribution of heat transfer coefficient, Schlieren optical technique was used to reproduce thermal field, and it was found that the induced flow creates thermally inactive regions just upstream and downstream of each protruding element.

Experimental and numerical studies were conducted by Shakerin et al. [17] to find out the flow behavior of air over a heated wall with single and repeated, two-dimensional, rectangular roughness elements. The flow visualization studies confirm the formation of nearly stagnant regions between the ribs and surface heat flux in these regions was very low. So the presence of wall ribs did not contribute to heat transfer enhancement. Zhongqi et al. [18] used Volume of Fluid (VOF) method to investigate the depressurization and thermal stratification behavior of a liquid nitrogen tank with different baffle structures under microgravity conditions. By optimizing the baffle setting, a reduction of up to 54% in pressurization rate was achieved, which is an eye-catching improvement for extended duration missions. Justin Oliveira et al. [19] investigated the effect of isogrid on thermal stratification inside propellant tanks. Studies showed that the boundary layer thickness on the wall in a forced free stream flow was distinctly thicker (150–700%) than the equivalent flat plate boundary layer thickness. Isogrids can either enhance or suppress stratification rate compared to smooth tanks depending upon roughness size and tank conditions. Experimental and numerical studies were carried out by Faure et al. [20] to assess the boundary layer behavior over the propellant tank with mass saving isogrid structures. They revealed that more than 200% thicker velocity boundary layer is developed over isogrid wall than a smooth wall. It leads to rapid self-pressurization and enhanced fluid mixing.

The presence of roughness elements has been found to reduce thermal stratification. The effect of spacing to height ratios of transverse ribs, protrusion length, the conductivity of ribs on heat transfer performance has been studied extensively, but for flat vertical and horizontal heated plates. Influence of ribs and grid structure on the reduction of stratification on a cryogenic cylindrical tank demand more detailed investigation.

3. Numerical modeling

A numerical model using Ansys Fluent is developed to study the stratification and self-pressurization phenomenon in a cryogenic storage tank. Smooth wall condition is considered and VOF (Volume of Fluid) transient method is used for the simulation. The governing equations are:

Conservation of mass:

$$\frac{\partial \rho}{\partial t} + \nabla.\left(\rho \vec{V}\right) = 0 \tag{1}$$

Conservation of momentum:

$$\frac{\partial(\rho \vec{V})}{\partial t} + \nabla.\left(\rho \vec{V}\vec{V}\right) = -\nabla P - \rho\beta\vec{g}(T - T_0) + \nabla.\left[\mu_{eff}\left(\nabla\vec{V} + \nabla\vec{V}^{T}\right)\right] + \vec{F} \tag{2}$$

Conservation of energy:

$$\frac{\partial(\rho CT)}{\partial t} + \nabla.\left(\vec{V}(\rho CT + P)\right) = \nabla.(K_{eff}\nabla T) + S_h \tag{3}$$

The location of the interface is identified by solving the continuity equation for the volume fraction of the second phase.

$$\frac{\partial}{\partial t}(\alpha_v\rho_v) + \nabla.\left(\alpha_v\rho_v\vec{V}\right) = \dot{m} \tag{4}$$

Where m represents the phase change mass at the interface due to evaporation or condensation. The density, viscosity and thermal conductivity is defined in terms of volume fractions.

$$\text{Density}, \rho = \alpha_l\rho_l + \alpha_v\rho_v \tag{5}$$

$$\text{Viscosity}, \mu = \alpha_l\mu_l + \alpha_v\mu_v \tag{6}$$

$$\text{Thermal conductivity}, k = \alpha_l k_l + \alpha_v k_{v.} \tag{7}$$

The assumptions used are,

1. The ullage pressure is determined by the saturation value corresponding to liquid–vapor interface.

2. The atmospheric temperature is considered as steady.

3. The ullage pressure is assumed to be uniform throughout.

Boundary conditions:

On the side walls,

$$-K\frac{\partial T}{\partial n} = q_w \tag{8}$$

At the bottom and top surfaces, the walls are adiabatic,

$$\frac{\partial T}{\partial n} = 0 \tag{9}$$

Where n is the direction normal to the wall

3.1 Numerical implementation

A cylindrical tank with 0.5 m diameter and 1 m height is used for the studies. The wall thickness is considered as 0.003 m and 2-D geometry is considered. Commercial CFD package Ansys 15 is used for solving the conservation equations. Axis-symmetric condition is selected because of the nature of physics, geometry and boundary conditions. The Rayleigh number corresponds to all operating condition is above the critical value and flow is always turbulent. So, k-ϵ turbulence model with enhanced wall function approach is applied. A constant heat flux of 10 W/m^2 is applied on the left sidewall. The pressure–velocity coupling algorithm selected is SIMPLEC (Semi-Implicit Method for Pressure-Linked Equation- Consistent). The converged solution is easy to achieve by using this method than a SIMPLE algorithm. The body forced weighted average scheme is used for solving the momentum equation. For tracking the liquid-vapor interface, the Geometric Reconstruction Scheme is applied. Since the problem is transient, a time step of 0.001 s is selected so that the Courant number is less than 0.1.

3.2 Grid independency

Three grid systems with mesh numbers 15288, 22893 and 30671 are used to test the resolution. **Table 1** shows the variation of pressure inside the tank with mesh number for the time period of 100 s. The self-pressurization for the two grids (22893 and 30671) is almost the same as the maximum pressure difference is less than 165 kPa. For the mesh number of 30671, the solution is time-consuming. The grids with 22893 quadrilateral grid elements are used with successively increasing mesh numbers towards the wall is selected for the present work.

Time (s)	Tank Pressure (Pa)		
	Grid no: 15288	Grid no: 22893	Grid no: 30671
10	101642	101690	101720
20	102057	102027	102030
30	102355	102392	102442
40	102610	102765	102838
50	102986	103134	103240
60	103214	103426	103555
70	103456	103640	103803
80	103614	103801	103929
90	103875	104060	104171
100	104012	104350	104483

Table 1.
Effect of mesh size on self-pressurization.

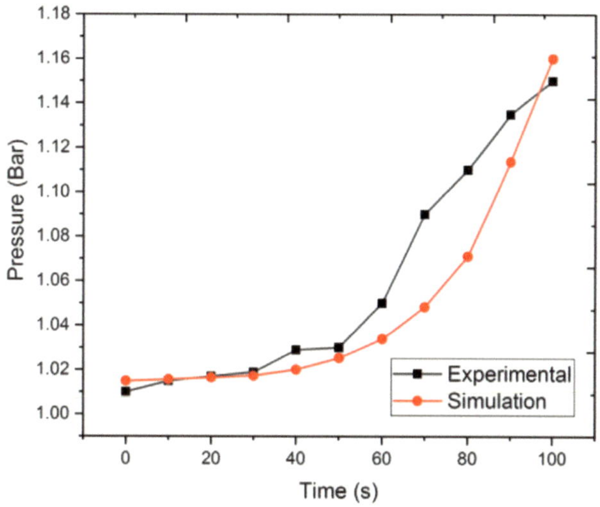

Figure 2.
Validation of the numerical model with experimental result.

3.3 Validation of the model

The numerical model developed has to be validated with the experimental results to prove the validity of the model in self-pressurization studies. The numerical model is validated with experiments conducted by Vishnu et al. [21]. The experiment set up consists of cylindrical test tank with a diameter of 0.11 m and height of 1.4 m. The liquid nitrogen cryogen was used as the model propellant and fill height was 0.7 m. The overall thermal conductivity of the tank wall was calculated as 0.01957 W/m.K for a tank thickness of 0.15378 m. Applying tank wall temperature of 300 K gives equivalent heat flux as that of experimental condition. The initial liquid temperature selected is 79.1 K which is equal to the initial interface temperature corresponds to the experimental condition and initial pressure corresponds to the saturation value of interface temperature. The initial temperature is assumed to be same throughout vapor and liquid domain. **Figure 2** compare the predicted pressure evolution against the experimental data. The maximum under-prediction of tank pressure by the numerical model compared to the experimental pressure data is not more than 2.6 percent. The tank numerical pressurization rate is in good agreement with the experiment, and hence self-pressurization is satisfactorily predicted.

4. Effect of obstruction elements

This section discusses about the impact of stratification due to the presence of roughness elements on the propellant tank wall. The nature of flow through a tank wall with roughness vary significantly to that of a smooth wall tank. The presence of roughness elements affects the formation of velocity and thermal boundary layer. The natural convection flow over a roughness element is more analogous to flow behavior over large scale obstruction elements such as forward-facing steps, backward facing steps, ribs and fences [20]. Based on the geometry, tank surface with obstruction elements can be classified as grids and ribs.

4.1 Comparison between flow over a smooth and rough wall

While comparing the performance of tanks with and without obstruction elements, proper scaling of tank geometry should be considered. Thermal stratification of a smooth wall with same liquid filling height as run length can be compared with that of a rough wall tank, but the volume of liquid and heated surface area will not be identical. Similarly, the heat flux applied could be matched but the filling height or run length would not be identical. In this analysis, the liquid filling height and tank radius considered are similar between smooth and rough wall tanks. Both tanks are having a radius of 0.25 m and filling height 0.5 m which is filled with liquid hydrogen. An axial heat flux of 10 W/m^2 is applied on the side wall. There are 25 number of obstruction elements with 0.178 cm height and 0.076 cm thickness. The details are shown in **Table 2**. The total run length along the obstruction surface is 108.9 cm which is 8.9% more than that of smooth wall tank. The smooth wall tank has a volume of 196349.54 cm^3 whereas rough wall tank has 196286.87 cm^3 which is 0.0319% lesser than smooth wall. The total heated surface area of rough wall tank is 17105.008 cm^2 which is 8.90% more than smooth wall tank and hence the volume to surface area ratio becomes 8.189% lesser for rough wall tank. The tank geometry and obstruction element layout is shown in **Figure 3**.

4.2 Comparison of flow through a smooth and rough surface tank

To study the effect of presence of obstruction elements on the natural convection flow, two cases were simulated. Flow through a smooth wall tank and flow through a storage tank with roughness elements at the inner wall surface. **Figure 4**

Parameter	Value
Tank height	1 m
Tank radius, R	0.25 m
Filling height, H	0.5 m
Grid height, h	0.178 cm
Grid thickness, t	0.076 cm

Table 2.
Geometrical parameters of tank.

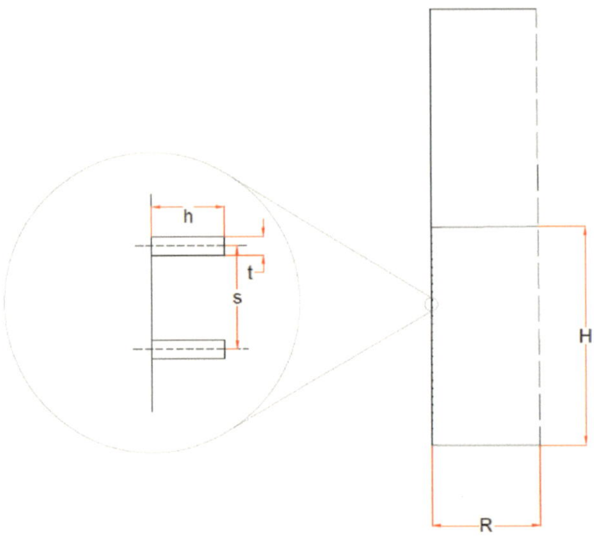

Figure 3.
Geometrical parameters of grids in a storage tank.

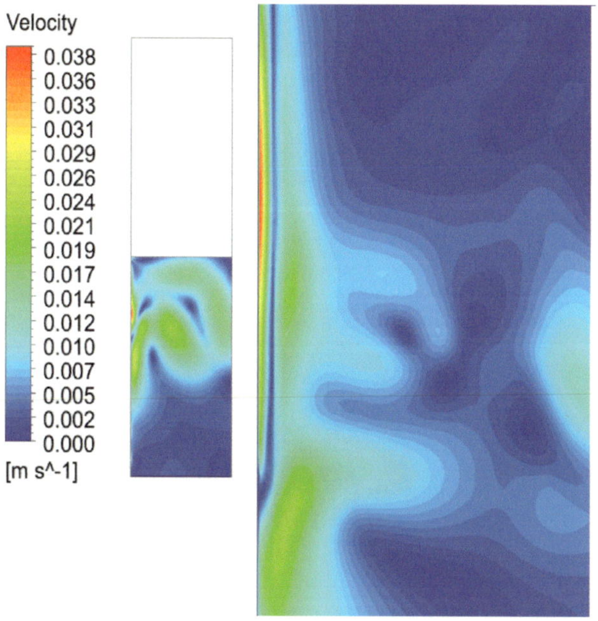

Figure 4.
Development of velocity boundary layer over a smooth surface tank.

shows the development of velocity boundary layer in a smooth tank wall due to natural convection flow. We can identify the development of laminar boundary layer and its magnitude keeps on increasing along the run length. The maximum velocity value of 0.038 m/s is obtained near to tank wall which is indicated as red color zone. Comparatively stagnant or undisturbed zones were developed at the bottom part of the bulk liquid zone.

Figure 5 shows the development of laminar boundary layer over a rough surface tank. Due to the presence of obstruction elements, we can see that the velocity near the tank wall is zero which is indicated by blue color zone. Also, stagnant regions are developed around the obstruction elements which causes further hindrance to the flow. Compared with smooth wall case, the bulk liquid seems undisturbed and flow due to natural convection takes place through the top face of the obstruction elements.

For the better understanding of the flow behavior over a tank surface with obstruction elements, stream line diagram can be used. The **Figure 6** shows the stream lines over a rough surface tank. It is clear that the flow gets obstructed with roughness elements. Stream line deflection also takes place due to the presence of obstruction elements. Apart from velocity distribution, the development of stratified layer greatly depends on the mechanism of heat transfer to the interface and bulk liquid.

Figure 7 shows the comparison of temperature contour over tank with smooth and roughness elements after 150 seconds of flow. The development of stratification and degree of stratification is higher for tank with obstruction elements. The presence of roughness elements causes the formation of stagnant regions near the wall but the increase in heat transfer area leads to better heat transfer to the fluid.

Figure 8 shows a detailed view of the development of thermal boundary layer over a tank with roughness elements. It can be seen that the boundary layer formed is almost uniform throughout the run length. It does not thicken monotonically along the wall. Due to increased surface area, there will be evident additional heating. So the temperature of liquid will be higher for rough wall case. The thermal

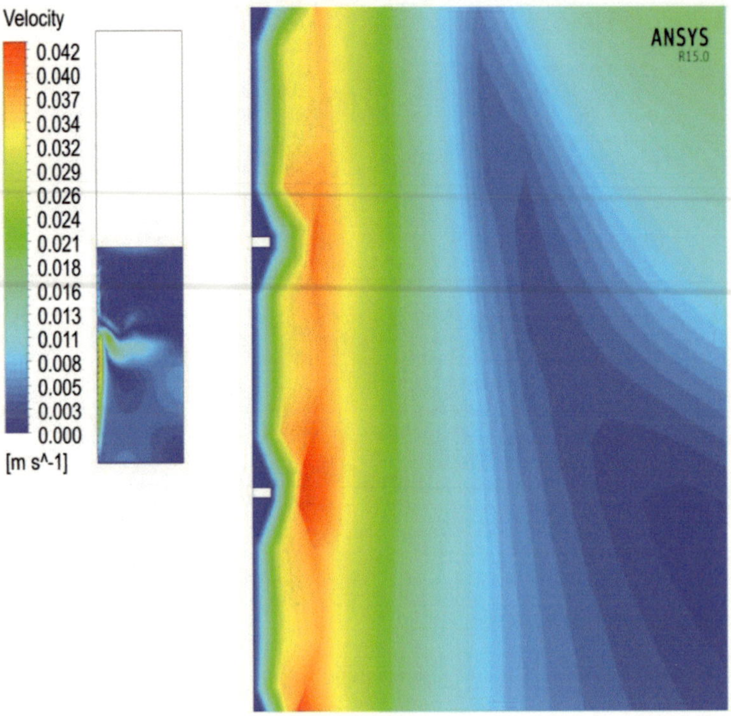

Figure 5.
Development of velocity boundary layer over a tank with obstruction elements.

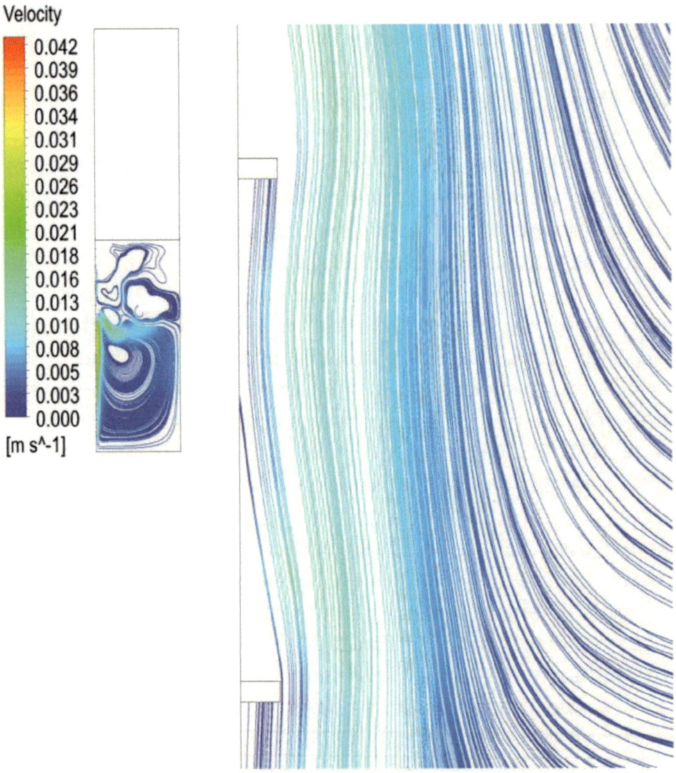

Figure 6.
Contour of stream lines over a tank with obstruction elements.

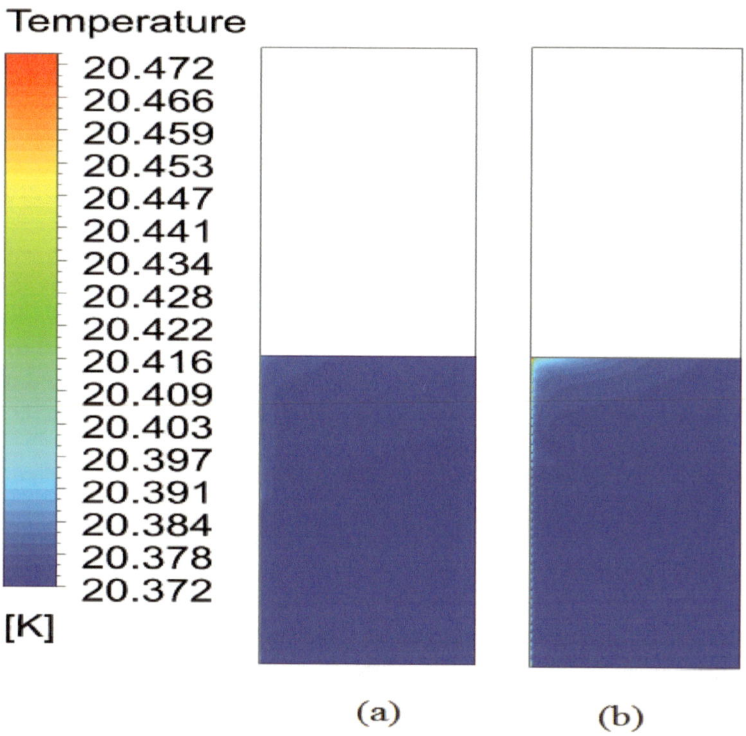

Figure 7.
Comparison of temperature contour over (a) smooth wall and (b) rough surface tank.

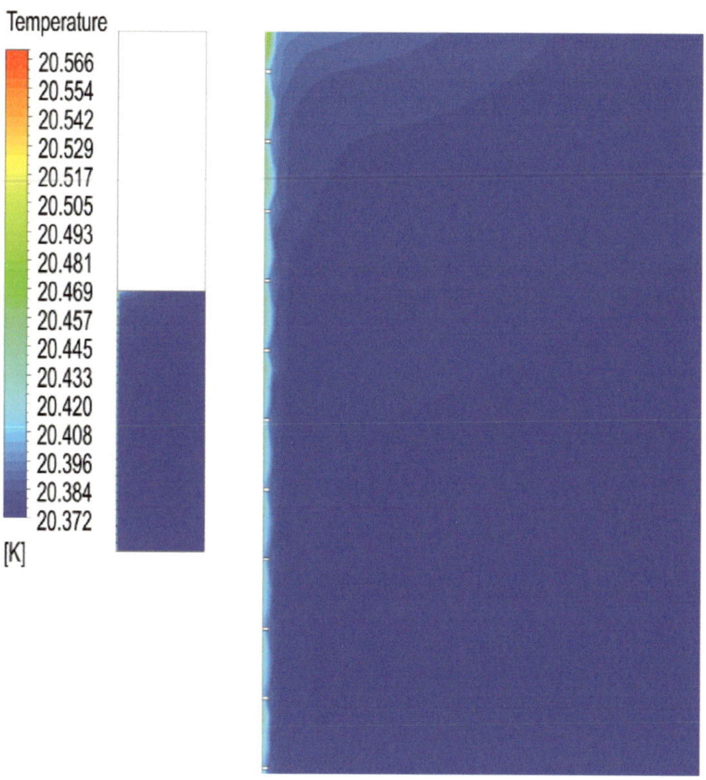

Figure 8.
Temperature contour of a liquid hydrogen tank with isogrids on the inner surface.

boundary layer developed is also thicker for rough wall case which further results in increased rate of thermal stratification and self-pressurization rate.

4.3 Thermal stratification in a cryogenic storage tank with isogrid surface

To study the impact on thermal stratification due to the presence of grids on tank wall, three cases were simulated. The roughness elements number varies such as 25, 30 and 35 corresponds to case 1, 2 and 3 respectively. For smooth wall case, the total volume of the tank is 196349.54 cm^3, the heated surface area is 15707 cm^2 and hence the volume to surface area ratio is 12.499. For tank with grids (case 1), the heated surface area increased to 17105.008 cm^2 and volume reduced to 196286.87 cm^2. So, the volume to surface area ratio reduces to 11.4754 which is 8.189%. Similarly, there is a reduction in volume to surface area ratio of 9.667% and 11.112% corresponds to case 2 and 3.

The evolution of pressure under three different cases is shown in **Figure 9**. As the number of elements increases, the value of pressure inside the tank also increases. The rise in pressure is noticeable after a time period of 60 seconds. It may due to the initial transient boundary layer formation process. As the number of roughness elements increases from 25 to 35, the heated surface area increases by 8.89% and 11.11%. It causes an increase in self-pressurization rate of 1.88 Pa/s (smooth wall) to 2.21 Pa/s (case 3). The case 3 with 35 number of obstruction elements have a self-pressurization rate which is 17.31% more than that of smooth wall case. The increase in self-pressurization rate due to presence of grids is shown in **Table 3**.

4.4 Thermal stratification in a cryogenic storage tank with rib surface

A similar kind of analysis is done incorporating ribs instead of grids on the tank wall. The rib geometry used in the analysis is having height 'h', spacing 's' and

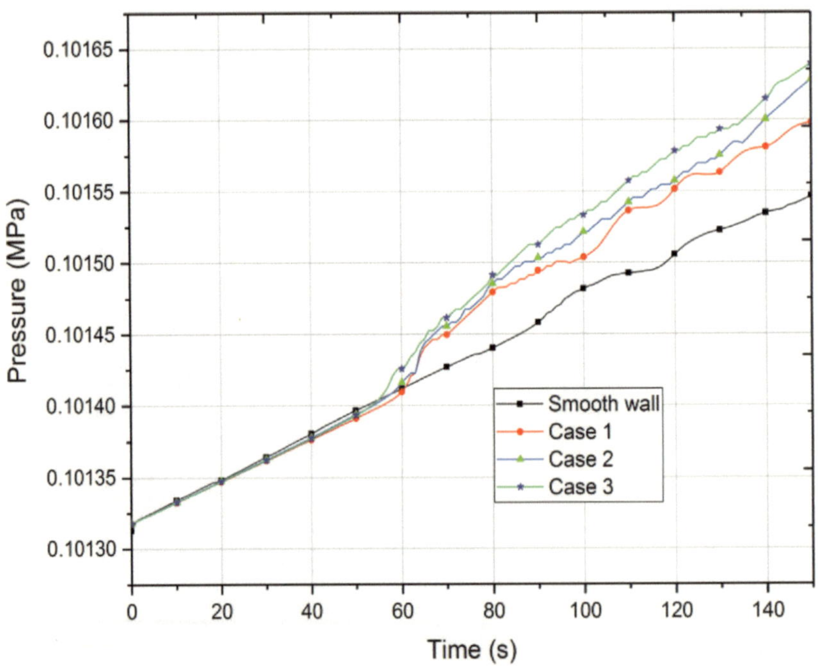

Figure 9.
Comparison of pressure evolution of a tank with different number of isogrids.

	Smooth wall tank	Case 1	Case2	Case3
Reduction in volume (%)	—	0.0319	0.0324	0.03787
Increase in heated surface area (%)	—	8.90	10.68	12.46
Reduction in VSA (%)	—	8.189	9.667	11.112
Increase in run length (%)		8.9	10.68	12.46
Self pressurization rate (Pa/s)	1.8866	1.92	2.1466	2.2133
Increase in SPR (%)	—	1.802	13.78	17.31

Table 3.
Influence of grid structure on the evolution of self-pressurization.

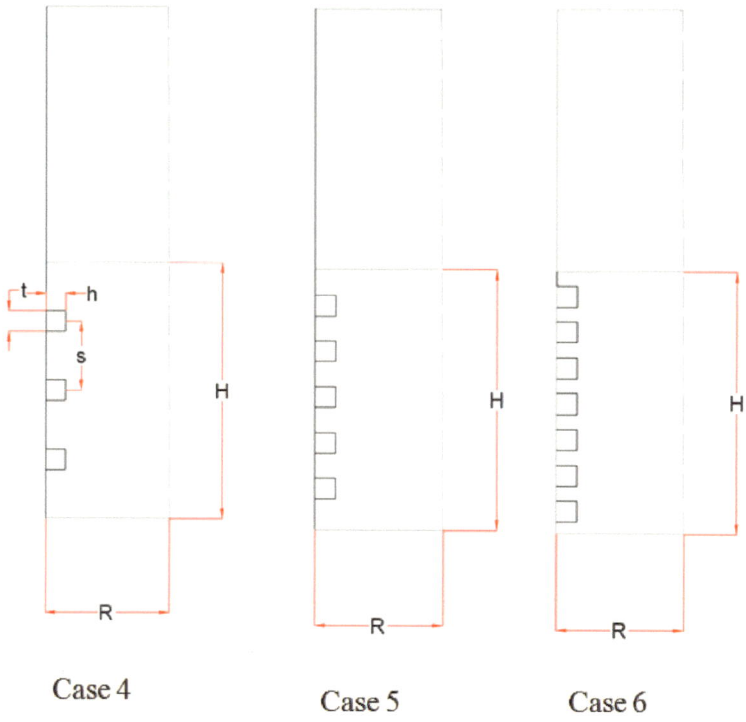

Figure 10.
Geometrical parameters of cylindrical tank with rib structure.

thickness 't' with rectangular cross section. The ribs were provided in the liquid side only. To study the effect of ribs on the evolution of stratification, three cases were considered; case 4, 5 and 6. Keeping the rib geometry and dimensions same, the distance between them is varied. So, the number of ribs corresponds to the cases 4 to 6 will be 3, 5 and 7 respectively. The tank geometry and rib layout are shown in **Figure 10**. The geometrical parameters are shown in **Table 4**.

Case	h (cm)	t (cm)	s (cm)
4	5	5	13.5
5	5	5	9
6	5	5	6.75

Table 4.
Geometrical parameters of ribs corresponds to cases 4–6.

For smooth wall case, the volume of the tank is 196349.54 cm^3 the heated surface area is 15707 cm^2 and hence the volume to surface area ratio is 12.499. For tank with 3 ribs (case 4), the volume reduces to 188809.7 cm^3 and heated surface area increases to 19476.91 cm^2. So, the volume to surface area ratio reduces to 9.694 which is 22.411%. Similarly, there is a reduction in volume to surface area ratio of 33.13% and 41.63% corresponds to case 5 and 6. Apart from these statistics, there is more increase in run length due to the presence of ribs. For case 4, the run length increases by 30%, 50% and 70% corresponds to case 5 and 6.

4.5 Effect of ribs on stratification

Figure 11 shows the velocity vector diagram for a tank with three grids (case 4). It is evident that the ribs create disturbance in the flow field, creating wake regions in the space between successive ribs. It affects the formation of boundary layer throughout the tank wall surface. Ultimately the presence of ribs enhances mixing in the bulk liquid which may reduce the formation of stratified layer.

The nature of flow velocities over three different configurations is depicted in **Figure 12**. The number of ribs affects the flow velocity inside the tank. Case 6 is having maximum number of ribs (7) which possess less velocity among the three cases. More number of ribs results in less flow velocity which indicates better mixing of bulk fluid and delayed stratification.

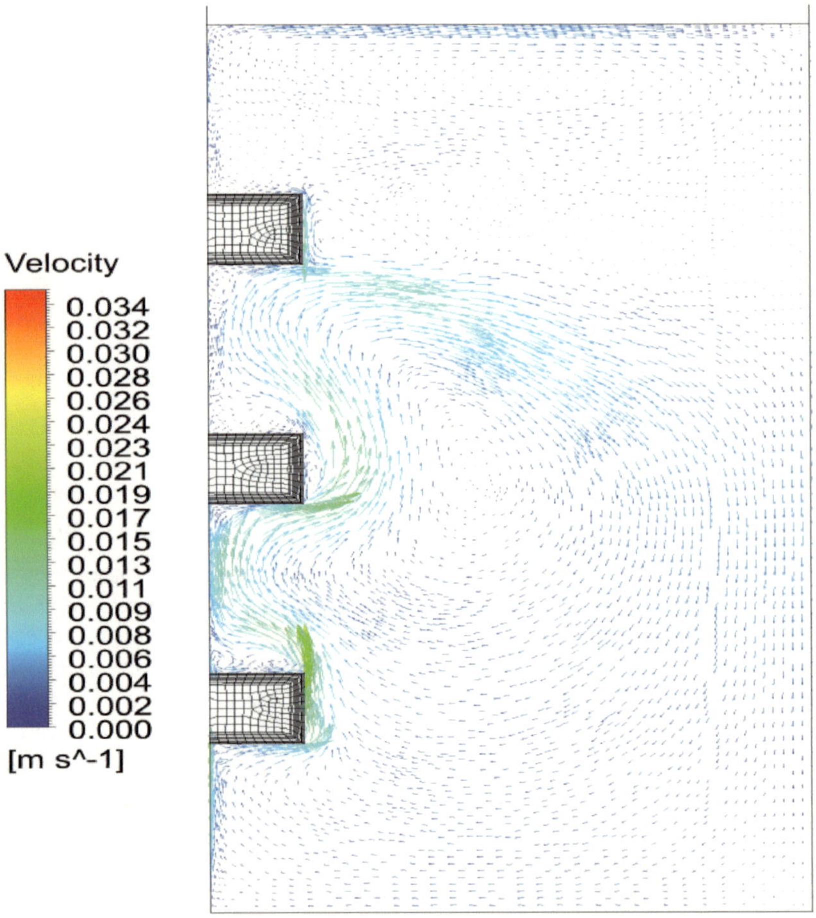

Figure 11.
Velocity vector diagram of a tank with rib structure.

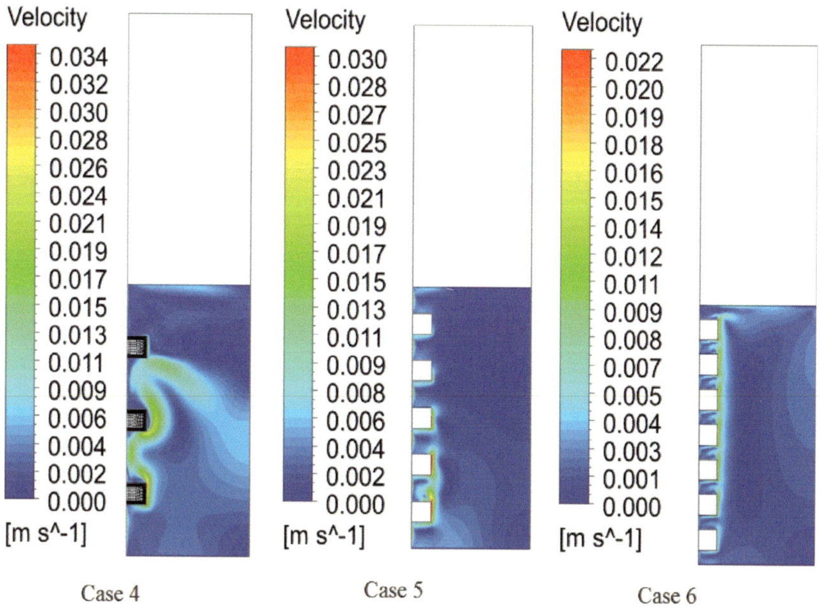

Figure 12.
Comparison of velocity of flow for cases 4–6.

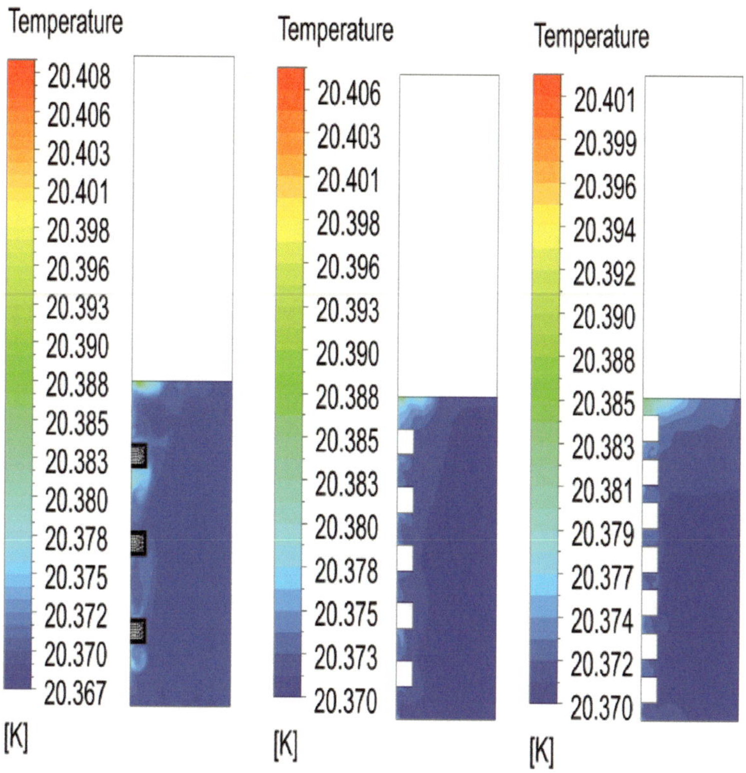

Figure 13.
Comparison of temperature of liquid for cases 4–6.

Figure 13 shows the comparison of temperature over three cases. Apart from interface, local heating zones were created on both sides of the rib surface. Since the presence of ribs causes obstruction of flow to a large extend, the locally heated fluid cannot travel to the interface effectively. The presence of recirculation zones enhances the mixing of the fluid. As a result, stratification develops very slowly. The formation of stratified layer is entirely different for three cases. There is more resembles between case 4 and smooth wall case. As the number of ribs increases, formation of local hotspots adjacent to the rib wall were visible and it is not transported to the interface. **Figure 14** shows the volume fraction under these cases.

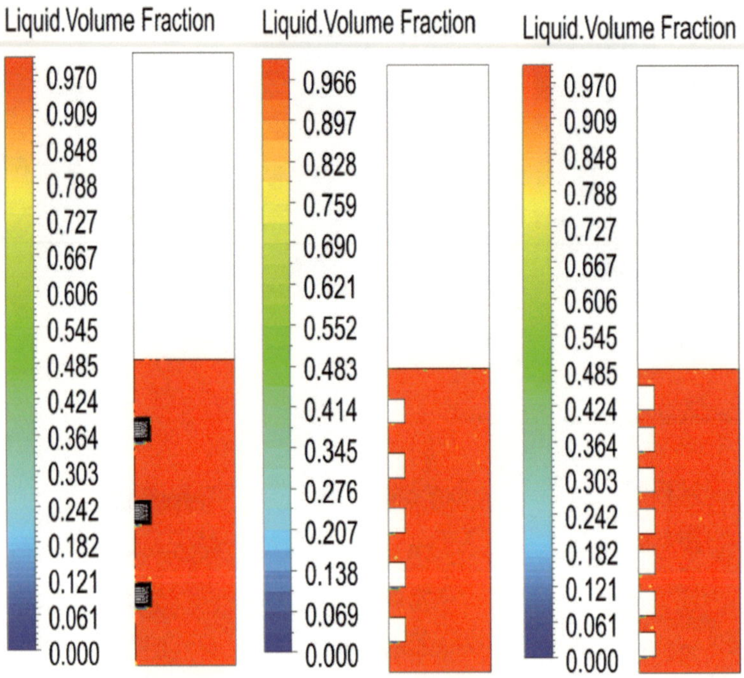

Figure 14.
Comparison of liquid volume fraction for cases 4–6.

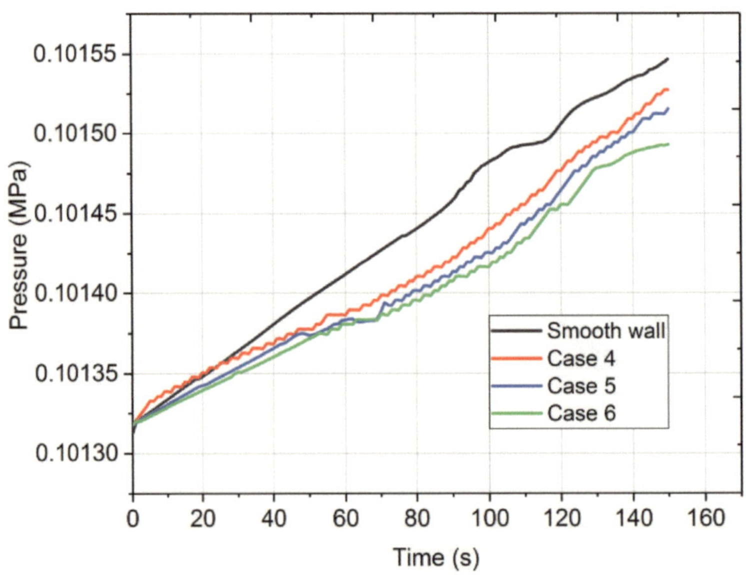

Figure 15.
Comparison of pressure evolution of a tank with different number of rib elements.

	Smooth wall tank	Case 4	Case5	Case6
Reduction in volume (%)	—	3.84	6.399	8.959
Increase in heated surface area (%)	—	24	40	56
Reduction in VSA (%)	—	22.441	33.13	41.63
Increase in run length (%)		30	50	70
Self pressurization rate (Pa/s)	1.8866	1.44	1.3533	1.266
Decrease in SPR (%)	—	23.67	28.26	32.89

Table 5.
Influence of rib structure on the evolution of self-pressurization.

Similar to temperature profile, the tank with more number of ribs has less phase change.

The evolution of pressure under three different cases is shown in **Figure 15**. As the number of ribs increases, the value of pressure inside the tank decreases and its value is well below that of smooth wall tank. The major change in pressure is noticeable after a time period of 70 seconds. As the number of grids increases from 3 to 7, the heated surface area increases by 24% and 56%. At the same time, a maximum reduction of 41.63% in volume to surface area ratio occurs for case 6. Ultimately the self-pressurization rate decreases by 32.89% for case 6. The decrease in self-pressurization rate due to presence of ribs is shown in **Table 5**.

5. Conclusions

A numerical model is developed to understand the effect of surface roughness elements on the evolution of stratification and self pressurization. The Volume of Fluid (VOF) method is used to predict the liquid–vapor interface movement, and the Lee phase change model is adopted for evaporation and condensation calculations. From previous studies, it is observed that the shape of the inner wall surface of the storage tank plays an essential role in the development of the stratified layer. Using the computational model developed, a detailed study has been conducted on a cylindrical storage tank with an iso grid and rib structure. The major conclusions are:

- The buoyancy-driven free convection flow over iso grid structure result in velocity and temperature profile that differs significantly from a smooth wall case.
- The thermal boundary layer was always more significant for iso grid type obstruction, and these obstructions induce streamline deflection and recirculation zones, which enhances heat transfer to bulk liquid.
- A larger self-pressurization rate is observed for tanks with an iso grid structure.
- The presence of ribs results in the reduction of upward buoyancy flow near the tank surface, whereas streamline deflection and recirculation zones were also perceptible.
- A maximum reduction of 32.89% in the self-pressurization rate is achieved with the incorporation of rib structure in the tank wall.

Nomenclature

C	Specific heat, J/kg.K
D	Diameter of the tank, m
F	Body force, N/m3
g	Acceleration due to gravity, m/s^2
H	Initial liquid height, m
K	Thermal conductivity, W/(m.K)
L	Total height of the vessel, m
$\dot{m}$	Mass flow rate, kg/s
P	Pressure, Pa
$\dot{q}$	Heat flux, W/m^2
S_h	Energy source term
T	Temperature, K
t	Time, s
$\vec{V}$	Velocity vector, m/s

Greek symbols

α	Volume fraction
μ	Absolute viscosity, Pa.s
ρ	Density, kg/m^3

Subscripts

h	Heat
l	Liquid
v	Vapor

Author details

S.B. Vishnu* and Biju T. Kuzhiveli
Centre for Advanced Studies in Cryogenics, National Institute of Technology Calicut, Calicut, India

*Address all correspondence to: vishnusb90@gmail.com

References

[1] J. W. Tatom, W. H. Brown, L. H. Knight, and E. F. Coxe, "Analysis of Thermal Stratification of Liquid Hydrogen in Rocket Propellant Tanks," *Adv. Cryog. Eng.*, pp. 265–272, 1964, doi: 10.1007/978-1-4757-0525-6_31.

[2] A. F. Schmidt, J. R. Purcell, W. A. Wilson, and R. V Smith, "An Experimental Study Concerning the Pressurization and Stratification of Liquid Hydrogen 1, 2," *J. Res. Natl. Bur. Stand. Eng. Instrum.*, vol. 65, no. 2, pp. 81–87, 1961.

[3] J. M. Ruder, "Stratification in a pressurized container with sidewall heating," *AIAA J.*, vol. 2, no. 1, pp. 135–137, 1964, doi: 10.2514/3.2242.

[4] J. C. Aydelott, "Normal gravity self-pressurization of 9-inch- (23 CM) diameter spherical liquid hydrogen tankage.," 1967.

[5] H. Ji, J. Hochstein, and S. Schwartz, "Experimental Verification of Scaling Parameters for Thermal Stratification," *AIAA Aerosp. Sci. Meet.*, 1990.

[6] M. M. Hasan, C. S. Lin, N. T. Van Dresar, F. Park, and C. S. L. Hasan, Mohammad M, "Self-Pressurization of a Flightweight Liquid Hydrogen Storage Tank Subjected to Low Heat Flux," *ASME/AIChE Natl. Heat Transfe Conf. Minneapolis, Minnesota, July 28-31,1991*, no. September, 1991.

[7] A. A. Yamaji, Rita Szijártó, "STUDY OF THERMAL STRATIFICATION AND MIXING USING PIV Bogdán Yamaji, Rita Szijártó, Attila Aszódi Institute of Nuclear Techniques, Budapest University of Technology and Economics."

[8] T. N. Veziroglu and J. W. Sheffield, "Analysis and Optimization of Thermal Stratification and Self-Pressurization Effects in Liquid Hydrogen Storage Systems — Part 2 : Model Results and Conclusions," vol. 115, no. September 1993, pp. 7–10, 2013.

[9] M. M. H. Chin-Shun Lin, "Numerical Investigation of the Thermal Stratification in Cryogenic Tanks Subjected to Wall Heat Flux," pp. 1–325, 2016, [Online]. Available: papers3://publication/uuid/40B251BA-DFAA-408F-AD41-E4BEBB70284E.

[10] X. Li, G. Xie, and R. Wang, "Experimental and numerical investigations of fluid flow and heat transfer in a cryogenic tank at loss of vacuum," *Heat Mass Transf. und Stoffuebertragung*, vol. 46, no. 4, pp. 395–404, 2010, doi: 10.1007/s00231-010-0583-9.

[11] Z. Liu, L. Wang, Y. Jin, and Y. Li, "Development of thermal stratification in a rotating cryogenic liquid hydrogen tank," *Int. J. Hydrogen Energy*, vol. 40, no. 43, pp. 15067–15077, 2015, doi: 10.1016/j.ijhydene.2015.09.039.

[12] M. Kang, J. Kim, H. You, and D. Chang, "Experimental investigation of thermal stratification in cryogenic tanks," *Exp. Therm. Fluid Sci.*, vol. 96, pp. 371–382, 2018, doi: 10.1016/j.expthermflusci.2017.12.017.

[13] J. Fu, B. Sunden, and X. Chen, "Influence of wall ribs on the thermal stratification and self-pressurization in a cryogenic liquid tank," *Appl. Therm. Eng.*, vol. 73, no. 2, pp. 1421–1431, 2014, doi: 10.1016/j.applthermaleng.2014.05.086.

[14] T. K. Khurana, B. V. S. S. S. Prasad, K. Ramamurthi, and S. S. Murthy, "Thermal stratification in ribbed liquid hydrogen storage tanks," *Int. J. Hydrogen Energy*, vol. 31, no. 15, pp. 2299–2309, 2006, doi: 10.1016/j.ijhydene.2006.02.032.

[15] G. Polidori and J. Padet, "Transient free convection flow on a vertical surface with an array of large-scale

roughness elements," *Exp. Therm. Fluid Sci.*, vol. 27, no. 3, pp. 251–260, 2003, doi: 10.1016/S0894-1777(02)00296-0.

[16] G. Tanda, "Natural convection heat transfer in vertical channels with and without transverse square ribs," *Int. J. Heat Mass Transf.*, vol. 40, no. 9, pp. 2173–2185, 1997, doi: 10.1016/S0017-9310(96)00246-3.

[17] A. Hadim and M. Ramot, "Natural convection in an enclosure with discrete heat sources on the vertical walls," *Am. Soc. Mech. Eng.*, vol. 31, no. 7, pp. 1–7, 1993.

[18] Z. Q. Zuo, W. B. Jiang, and Y. H. Huang, "Effect of baffles on pressurization and thermal stratification in cryogenic tanks under micro-gravity," *Cryogenics (Guildf).*, vol. 96, pp. 116–124, 2018, doi: 10.1016/j.cryogenics.2018.10.017.

[19] J. Oliveira, D. R. Kirk, S. Chintalapati, and P. A. Schallhorn, "Effect of isogrid roughness on thermal stratification," *Collect. Tech. Pap. - 43rd AIAA/ASME/SAE/ASEE Jt. Propuls. Conf.*, vol. 5, no. July, pp. 4734–4748, 2007, doi: 10.2514/6.2007-5497.

[20] J. M. Faure, J. M. Oliveira, S. Chintalapati, H. M. Gutierrez, and D. R. Kirk, "Effect of isogrid-type obstructions on thermal stratification in upper-stage rocket propellant tanks," *J. Spacecr. Rockets*, vol. 51, no. 5, pp. 1587–1602, 2014, doi: 10.2514/1.A32699.

[21] S. B. Vishnu, S. Bhowmick, and B. T. Kuzhiveli, "Experimental and Numerical Investigation of Stratification and Self pressurization in a High Pressure Liquid Nitrogen Storage Tank," *Energy Sources, Part A Recover. Util. Environ. Eff.*, vol. 0, no. 0, pp. 1–15, 2019, doi: 10.1080/15567036.2019.1651424.

Chapter 4

Ultimate Compressive Strength of Steel Stiffened-Plate Structures Triggered by Brittle Fracture under Cryogenic Conditions

Dong Hun Lee, Jeom Kee Paik, Jonas W. Ringsberg and P.J. Tan

Abstract

This chapter presents a practical method to investigate the effects of brittle fracture on the ultimate compressive strength of steel stiffened-plate structures under cryogenic conditions. Computational models are developed to analyse the ultimate compressive strength of steel stiffened-plate structures, triggered by brittle fracture, under cryogenic condition. A phenomenological form of the material model for the high-strength steel at cryogenic condition is proposed, that takes into account the Bauschinger effect, and implemented into a nonlinear finite element solver (LS-DYNA). Comparison between computational predictions and experimental measurements is made for the ultimate compressive strength response of a full-scale steel stiffened-plate structure, showing a good agreement between them.

Keywords: Bauschinger effect, brittle fracture, compressive strength, cryogenic condition, nonlinear finite element method, steel stiffened plate structure, ultimate strength

1. Introduction

Steel stiffened panels are used in naval, offshore, mechanical and civil engineering structures as their primary strength sub-structures. Occasionally, they are exposed to cryogenic conditions, e.g., due to the unwanted release of liquefied gas such as LNG (liquefied natural gas) or liquefied hydrogen as discussed in Paik et al. [1].

The ultimate strength is a primary criterion for the design of steel stiffened-plate structures [2–6], and it is essential to characterise the effects of cryogenic condition on the ultimate strength of such structures. The authors of this chapter have previously conducted collapse tests on full-scale steel stiffened-plate structures under axial-compressive loading at room temperature [1], at cryogenic condition [7] and at -80°C [8]. This chapter is part of a sequel to investigate the brittle fracture of steel stiffened-plate structures under cryogenic conditions. The tested structures were designed from a reference plate panel in an as-built containership carrying 1,900 TEU (twenty-foot equivalent units). They were fabricated at a shipyard using exactly the same welding technology as used in today's shipbuilding industry [9].

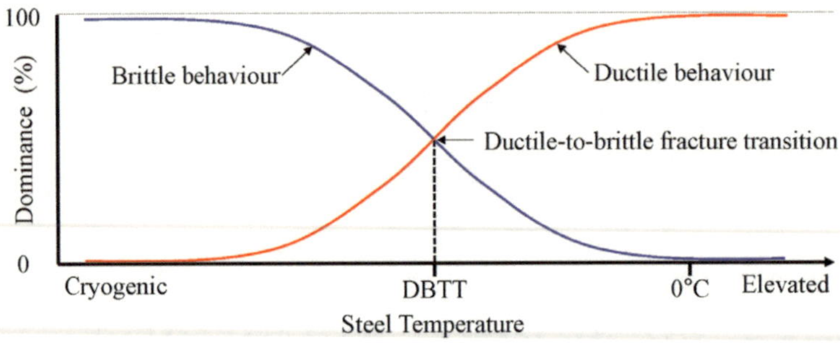

Figure 1.
Transition of the ductile-to-brittle behaviour for structural steel [5].

Structural fracture modes are classified into three groups [3, 4]: rupture, ductile fracture and brittle fracture. Material rupture occurs when failure occurs by cracking associated with necking localisation during large plastic flow. If the strain at which a material fractures is small, with very little ductility, it is a brittle fracture. An intermediate fracture mode between rupture and brittle fracture is called ductile fracture with partial ductility. Fracture behaviour of ductile materials such as carbon steels is quite different from that of inherently brittle materials. Ductile materials generally exhibit slow stable crack growth during crack extension, but they can show a similar behaviour to brittle materials at specific environments such as very low temperatures or lower than the ductile-to-brittle fracture transition temperatures (DBTT) and/or impact loading. It is also recognised that the Bauschinger effect of materials cannot be neglected at sub-zero temperatures and cryogenic condition as the material behaviour in compression is distinct from that in tension [3, 10, 11]. **Figure 1** illustrates the transition of the ductile-to-brittle behaviour for structural steel.

To compute the failure behaviour of structures at cold (sub-zero) temperatures (or higher than the ductile-to-brittle fracture transition temperatures), constitutive equations of materials have been proposed in the literature [12–15]. It is recognised that most of previous studies are associated with predominantly ductile behaviour or at least with partial ductility, but studies applicable to entirely brittle fracture at cryogenic condition are lacking. It is also recognised that the approaches using the constitutive equations are not always practical for the ultimate strength analysis of supersized structures because they are too complex to apply for the problem.

This chapter presents a method for computing the ultimate compressive strength of steel stiffened-plate structures by nonlinear finite element method (NLFEM) using the multi-physics software package LS-DYNA implicit code. Mechanical properties of high-strength steel with grade AH32 used for fabricating the tested structure were obtained from tension and compression tests at low temperatures and cryogenic condition [7, 8], and a phenomenological relation of engineering stress versus engineering strain of the material was formulated. The material model is implemented into the LS-DYNA implicit code. To demonstrate the validity of the computational model, the NLFEM is compared to experimental results from a full-scale physical test.

2. Literature survey on structural behaviour at cold (sub-zero) temperatures

A number of studies in modelling of material behaviour for structural steels at cold (sub-zero) temperatures are available in the literature. Most of the studies dealt with predominantly ductile behaviour of materials with the focus on how crack initiates in association with ductile fracture. Ehlers and Varsta [16] and Ehlers [17] derived the true stress versus true strain relation of ordinary steel. The effects of stress triaxiality on ductile fracture have been one of research topics [18–24]. The works of the Choung group have provided useful insights for ductile fracture behaviour of structural steels [14, 25–32].

It is recognised that structural steel behaviour is predominantly ductile at temperatures higher than the temperature of the ductile-to-brittle fracture transition, as shown in **Figure 1**. As the temperature decreases approaching cryogenic condition, the material behaves predominantly in a brittle manner with partial or no ductility [33–40]. Majzoobi et al. [41] observed that the ductile-to-brittle fracture transition of carbon steel occurs at about -80°C, and the material behaviour becomes entirely brittle at -196°C.

Although there are considerable uncertainties associated with the ductile-to-brittle fracture transition temperature (DBTT), a number of evidences for brittle fracture behaviour of steel structures at cryogenic condition have been seen in the literature, depending on the type of materials and loading conditions (e.g., quasi-static or impact), among other factors. Crushing testing of steel tubes under quasi-static loads at -60°C [42, 43] showed ductile fracture, as shown in **Figure 2**. Dropped-object impact testing of steel stiffened plate panels at -60°C [44] showed brittle fracture, as shown in **Figure 3**. Full-scale collapse testing of a steel stiffened-plate structure under axial-compressive loading showed that the ultimate strength was reached by a trigger of brittle fracture [1], as shown in **Figures 4** and **5**. At room temperature, the structures reached the ultimate strength by flexural-torsional buckling [1], but brittle fracture triggered the global failure at cryogenic conditions [7].

Here, an attempt is made to develop new fracture criteria based on the hypothesis that crack initiates if an equivalent stress exceeds a critical value, to model the fracture phenomenon of high-strength steel (AH32) under cryogenic conditions. Existing material models for the fracture analysis is first reviewed.

2.1 Maximum principal stress based fracture criterion

The maximum principal stress-based fracture criterion is the simplest among all fracture criteria. It is useful to predict fracture behaviour of brittle materials under predominantly tensile loads. In this criterion, brittle fracture is expected to occur when the largest principal normal stress reaches the ultimate tensile strength (σ_T) of the material, which is usually obtained from tension tests of coupon specimens. The maximum principal stress-based fracture criterion is expressed as follows:

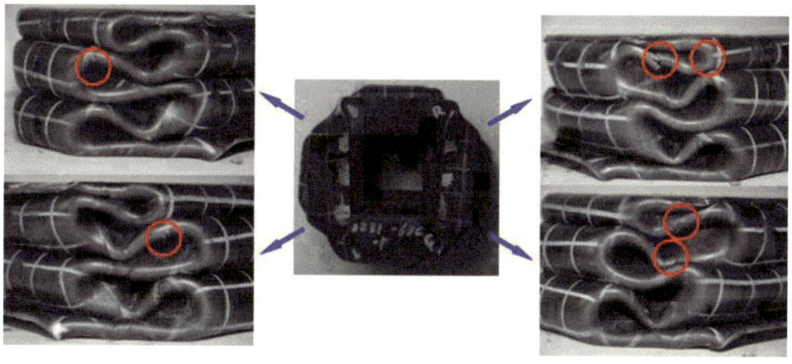

Figure 2.
Ductile fracture of a square tube under quasi-static crushing loads at -60°C [42].

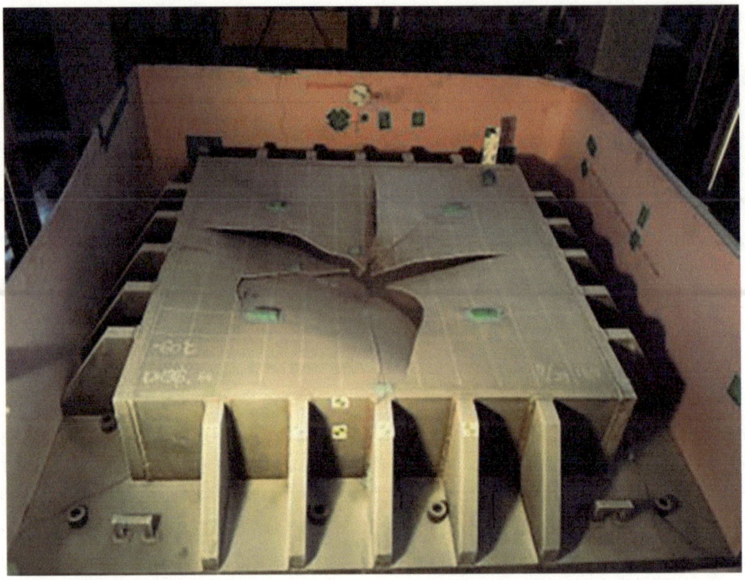

Figure 3.
Brittle fracture of a steel stiffened-panel under dropped-object impact at -60°C [44].

Figure 4.
The axial-compressive collapse test set-up of a full-scale steel stiffened plate structure [1, 7, 8].

$$Max.(|\sigma_1|, |\sigma_2|, |\sigma_3|) = \sigma_T \tag{1}$$

where σ_1, σ_2 and σ_3 are the principal stress components.

2.2 Coulomb-Mohr fracture criterion

The Coulomb-Mohr fracture criterion gives reasonably accurate predictions of fracture in brittle materials for which the compressive strength far exceeds the tensile strength, e.g., concrete or cast iron [45]. It is presumed that fracture occurs in a certain stress plane of material when a critical combination of normal stress and shear stress acts on the plane. The linear relation of the combination of critical stresses is given by:

Figure 5.
Brittle fracture of a full-scale steel stiffened plate structure under axial-compressive loads at -160°C [7].

$$|\tau| + \mu\sigma = \tau_i \tag{2}$$

where τ is the shear stress, σ is the normal stress, μ and τ_i are constants for a given material.

2.3 Johnson-Holmquist fracture criterion

The Johnson-Holmquist fracture criterion [46] is useful for modelling brittle materials, e.g., ceramic and glass, over a range of strain rates. It is one of the most widely used models in dealing with the ballistic impact on ceramics, which is expressible as follows:

$$\sigma^* = \sigma_i^* - D\left(\sigma_i^* - \sigma_f^*\right) \tag{3}$$

where σ_i^* is the uniaxial failure strength of intact material, see Eq. (4), σ_f^* is the uniaxial failure strength of completely fractured material, see Eq. (5), and D is a damage accumulation variable, see Eq. (6).

$$\sigma_i^* = A(p^* + T^*)^n \left[1 + C\ln\left(\frac{d\varepsilon_p}{dt}\right)\right] \tag{4}$$

$$\sigma_f^* = B(p^*)^m \left[1 + C\ln\left(\frac{d\varepsilon_p}{dt}\right)\right] \tag{5}$$

$$\frac{dD}{dt} = \frac{1}{\varepsilon_f}\frac{d\varepsilon_p}{dt} \tag{6}$$

where A, B, C, m, n are material constants, t is time, ε_p is the inelastic strain, and ε_f is the plastic strain to fracture. The asterisk indicates a normalised quantity, where the quantities of each variable are defined as follows:

$$\sigma^* = \frac{\sigma}{\sigma_{hel}}, p^* = \frac{p}{\sigma_{hel}}, T^* = \frac{T}{p_{hel}} \tag{7}$$

where σ^* and p^* are the stresses normalised by the stress at the Hugoniot elastic limit, and T^* is the tensile hydrostatic pressure normalised by the pressure at the Hugoniot elastic limit.

The Johnson-Holmquist damage model was modified by Deshpande and Evans [47] and Bhat et al. [48], where it is considered that the propagation of an initial crack is a function of the stress state, the fracture toughness and the flaw characteristics.

3. Proposed fracture criteria

The yield and ultimate tensile strengths of structural steels tend to increase with a decrease in the temperature [49], and subsequently the maximum load-carrying capacity (ultimate strength) of steel structures at cold temperatures is greater than that at room temperature [43, 50]. **Figure 6** shows a schematic of ductile and brittle failure behaviour, where the brittle fracture-induced ultimate strength P_{u2} at cryogenic condition is greater than the ductile collapse-induced P_{u1} at room temperature. However, the post-ultimate strength behaviour becomes very unstable if brittle fracture triggers the structural collapse at cryogenic condition. In this case, the strain energy absorption capability of structures can be more useful than the ultimate strength itself in terms of the structural safety assessment as it is obtained by integrating the area below the load–displacement curve until or after the ultimate strength is reached. The absorbed energy E_3 at cryogenic condition can be smaller than E_1 or E_2 at room temperature or a temperature higher than the DBTT (ductile-to-brittle fracture transition temperature). For this purpose, the entire

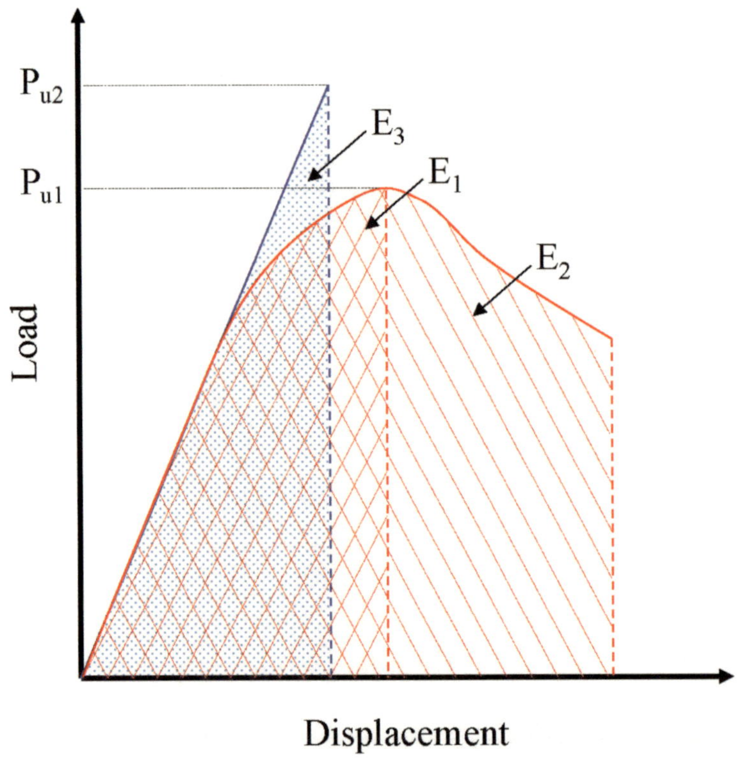

Figure 6.
Ultimate strength and post-ultimate strength behaviour at room temperature (or a temperature higher than the DBTT) versus cryogenic condition.

behaviour of structural collapse involving brittle fracture at cryogenic condition must be quantified efficiently and accurately.

3.1 Fracture criteria

A practical model is proposed for carbon steels which can be used for the ultimate strength analysis triggered by brittle fracture at cryogenic condition or in the region of ductile-to-brittle fracture transition. An elastic-perfectly plastic material model without the strain-hardening effect is used similar to a typical application at room temperature (20°C). However, the material behaviour in compression is different at low (sub-zero) temperatures or cryogenic condition from that in tension as the Bauschinger effect plays a role. However, the Bauschinger effect is usually neglected at room temperature with $\sigma_{YC} = \sigma_{YT}$, $\varepsilon_{Yc} = \varepsilon_{Yt}$ and $\varepsilon_{fc} = \varepsilon_{ft}$.

In the present model, it is hypothesised that brittle fracture occurs if the equivalent stress (σ_{eq}) reaches a fracture stress which is defined as the yield strength of material at the corresponding temperature, which can be expressed as follows:

(1) In tension:

$$\sigma_{eq} \geq \sigma_{FT} \text{ with } \sigma_{FT} = \gamma_t \sigma_{YT} \tag{8}$$

(2) In compression:

$$\sigma_{eq} \geq \sigma_{FC} \text{ with } \sigma_{FC} = \gamma_c \sigma_{YC} \tag{9}$$

where σ_{eq} is the equivalent stress, σ_{FT} and σ_{YT} are the fracture or yield stresses in tension at cryogenic condition or in the region of ductile-to-brittle fracture transition (which depends on types of materials), σ_{FC} and σ_{YC} are the fracture or yield stresses in compression at cryogenic condition or in the region of ductile-to-brittle fracture transition, γ_t and γ_c are test constants for a given steel in tension or compression, which may depend on various sources of parameters including chemical composition (grade), temperature and strain rate.

In Eqs. (6) and (7), σ_{eq} can be calculated as a function of principal stresses by the von Mises stress [51] as follows:

$$\sigma_{eq} = \frac{1}{\sqrt{2}} \sqrt{(\sigma_1 - \sigma_2)^2 + (\sigma_2 - \sigma_3)^2 + (\sigma_3 - \sigma_1)^2} \tag{10}$$

For plane stress state, σ_{eq} can be simplified as follows:

$$\sigma_{eq} = \frac{1}{\sqrt{2}} \sqrt{(\sigma_1 - \sigma_2)^2 + \sigma_2{}^2 + \sigma_1)^2} \tag{11}$$

3.2 Formulation of the engineering stress-engineering strain relations

The relations of the engineering stress versus engineering strain can be formulated following the fracture criterion defined in Section 3.1. If the steel temperature, T_s, is above the ductile-to-brittle fracture transition temperature (DBTT), i.e., $T_s >$ DBTT, the material behaves according to the ductile region. In this case, the stress–strain relation in tension is expressed as follows:

$$\sigma = \begin{cases} E\varepsilon \; for \; \sigma < \sigma_{YT} \\ \sigma_{YT} \; for \; \sigma = \sigma_{YT} \;\; or \; \varepsilon = \varepsilon_{Yt} \\ 0 \;\; for \; \varepsilon \geq \varepsilon_{ft} \end{cases} \tag{12}$$

where σ is the engineering stress, ε is the engineering strain, E is the elastic modulus, ε_{Yt} is the yield strain in tension, and ε_{ft} is the fracture strain in compression. **Figure 7** shows a schematic view of the proposed model of the engineering stress-engineering stress relation with full or partial ductility at room temperature or low temperatures.

In compression, the following equation similar to Eq. (13) is obtained.

$$\sigma = \begin{cases} E\varepsilon \ for\ \sigma < \sigma_{YC} \\ \sigma_{YC}\ for\ \sigma = \sigma_{YC}\ or\ \varepsilon = \varepsilon_{Yc} \\ 0\ \ for\ \varepsilon \geq \varepsilon_{fc} \end{cases} \tag{13}$$

where ε_{Yt} is the yield strain in tension and ε_{fc} is the fracture strain in compression.

If $T_s \leq$ DBTT or the steel temperature is lower than DBTT, the material behaves according to the brittle region. In this case, the stress–strain relation in tension is expressed as follows:

$$\sigma = \begin{cases} E\varepsilon \ \ for\ \sigma < \sigma_{FT} \\ 0 \ \ for\ \sigma \geq \sigma_{FT} \end{cases} \tag{14}$$

In compression, a similar equation to Eq. (14) is obtained as follows:

$$\sigma = \begin{cases} E\varepsilon \ \ for\ \sigma < \sigma_{FC} \\ 0 \ \ for\ \sigma \geq \sigma_{FC} \end{cases} \tag{15}$$

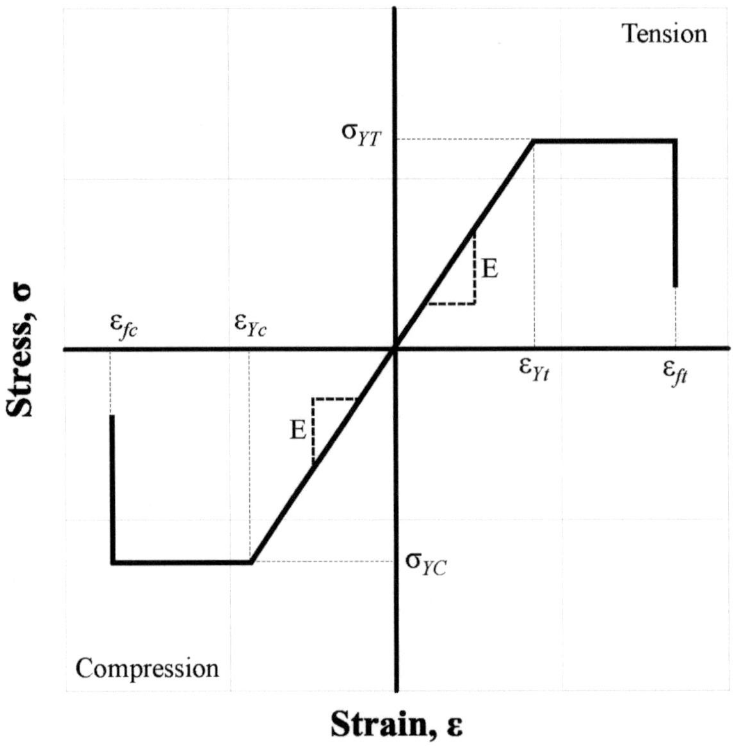

Figure 7.
Proposed model of the engineering stress-engineering stress relation with full or partial ductility at room temperature or sub-zero temperatures.

In the region of entire brittle fracture, i.e., with $T_s \leq$ DBTT, **Figure 7** can be redrawn as shown in **Figure 8** when elongation can be neglected after reaching the yield stress in tension or compression as the material exhibits brittle fracture immediately.

4. Nonlinear finite element method modelling

The proposed material model is implemented into the LS-DYNA implicit solver with 4-node shell elements (element formulation 16) for the ultimate compressive strength analysis of a structure tested at cryogenic condition as shown in **Figures 4** and **5** [7]. **Table 1** with **Figure 9** shows the dimensions of the tested structure. The nonlinear ultimate compressive strength analysis is simulated in a quasi-static analysis using the LS-DYNA implicit solver. The 4-node shell elements are used to model plating, stiffeners and transverse frames. In order to ensure sufficient resolution in the mesh size, a convergence study was performed by varying the element size following a typical approach as described in Paik [4]. The resulting convergence study provided the element size of 40 mm × 40 mm which was chosen to obtain sufficiently accurate results while minimising the computational cost. The thermal shrinkage effects for steel at low temperatures were not considered. **Figure 10** shows the FE model of the tested structure. **Figure 11** shows the loading and boundary conditions which were modelled as much as close to the tested structure, where unloaded edges were kept straight and loaded edges were entirely fixed.

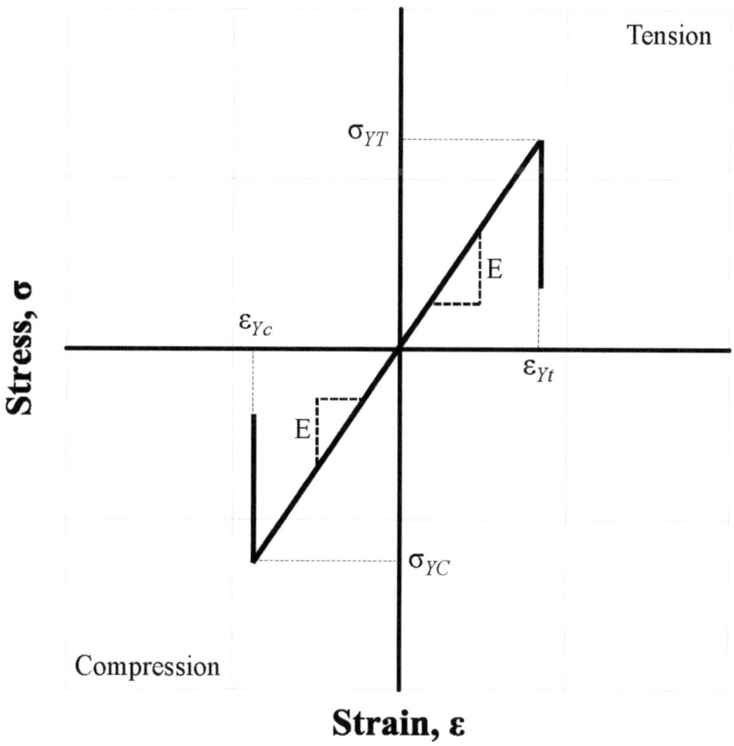

Figure 8.
Proposed model of the engineering stress-engineering strain relation without ductility or entire brittle behaviour below the ductile-to-brittle transition temperature or at cryogenic condition.

Material	AH32 high-strength steel
Spacing between transverse frames (*a*)	3150 mm
Spacing between longitudinal stiffeners (*b*)	720 mm
Plating thickness (*t*)	10 mm
Longitudinal stiffener (middle bay)	290 × 90 × 10/10 (T) (mm)
Longitudinal stiffener (side bays)	290 × 90 × 20/10 (T) (mm)
Transverse frame	665 × 150 × 10/10 (T) (mm)

Table 1.
Dimensions of the tested structure.

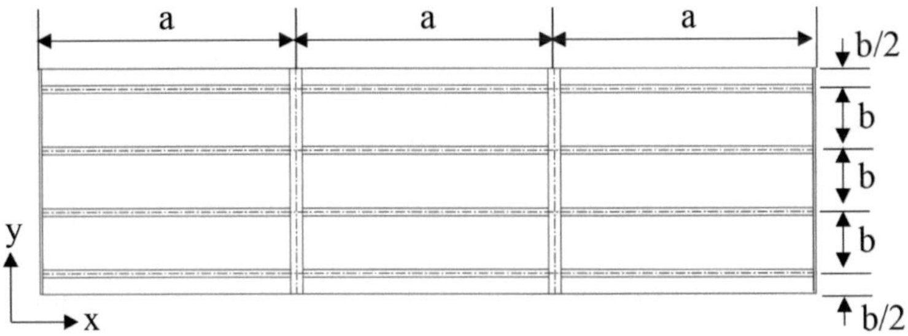

Figure 9.
Nomenclature of the scantlings for the tested structure.

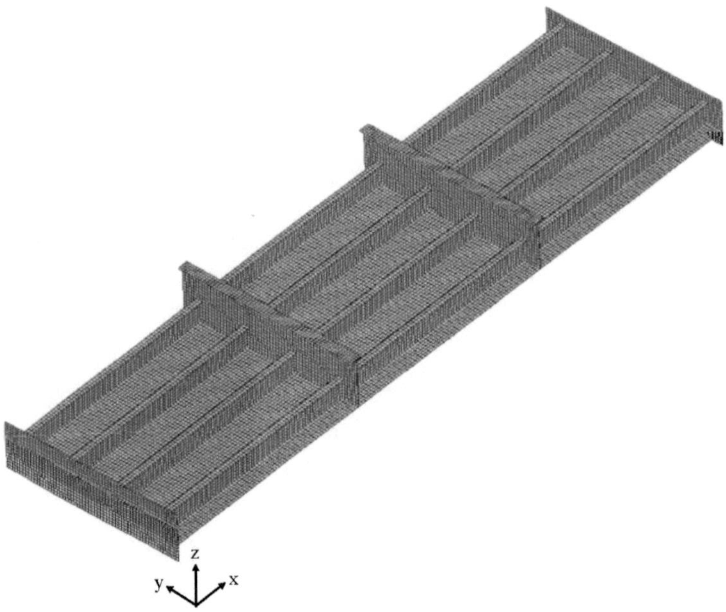

Figure 10.
Finite element mesh model of the tested structure.

Only the middle bay of the tested structure was exposed to the cryogenic condition as shown in **Figure 12**. **Table 2** summarises the measured data of steel temperatures during the collapse testing. For details of **Table 2**, see Paik et al. [7].

In room temperature, the mechanical properties of steel in compression are typically defined in the same manner as in tension without considering. However,

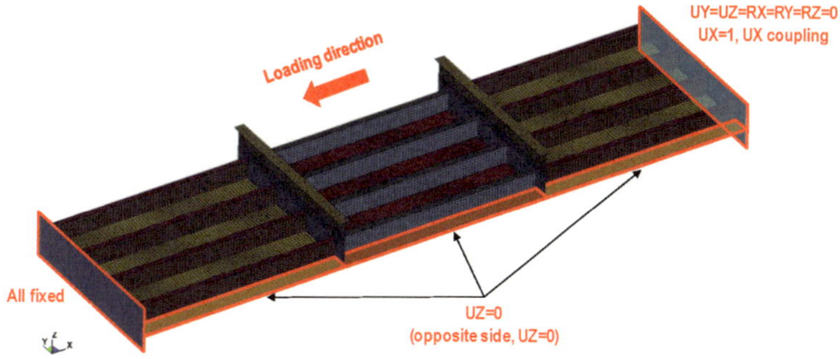

Figure 11.
The loading and boundary conditions applied to the FE model.

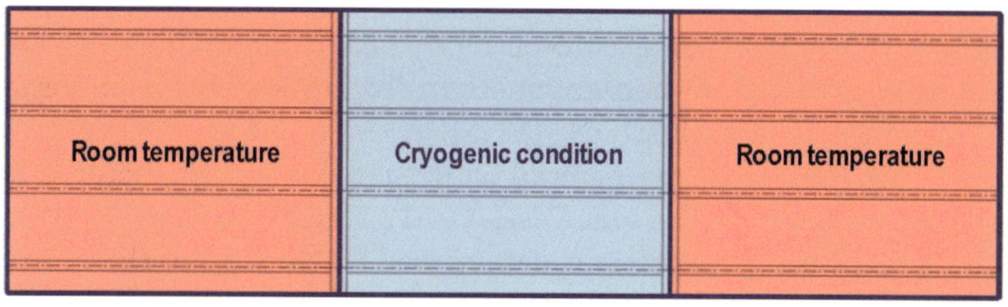

Figure 12.
Middle bay of the structure exposed to cryogenic condition.

Part	Highest temperature (°C)	Lowest temperature (°C)	Average temperature (°C)
Plating	−147.1	−175.4	−161.6
Web of longitudinal stiffener	−72.1	−167.2	−128.8
Flange of longitudinal stiffener	−58.8	−99.1	−79.3

Table 2.
Measured steel temperatures of the tested structure [7].

the Bauschinger effect cannot be neglected at low (sub-zero) temperatures and cryogenic condition [3, 10, 11]. To define the mechanical properties of AH32 steel at different temperatures (20°C, −80°C, −130°C and − 160°C), material tests in tension and compression were conducted. Details of these test data are presented in separate papers [7, 8]. **Tables 3** and **4** summarize the test data for the mechanical properties of the AH32 steel. It is found that the yield stress of steel in tension or compression increases as the temperature decreases, while the fracture strain in tension decreases with decrease in the temperature. The elastic modulus of steel

Parameter	At 20°C	At −80°C	At −130°C	At −160°C
Elastic modulus, E (GPa)	205.8	205.8	205.8	205.8
Yield stress, σ_{YT} (MPa)	358.0	433.4	546.7	672.9
Fracture strain, ε_{ft} (−)	0.376	0.430	0.409	0.336
Poisson's ratio (−)	0.3	0.3	0.3	0.3

Table 3.
Mechanical properties of AH32 steel at room and low temperatures in tension [7, 8].

Parameter	At 20°C	At −80°C	At −130°C	At −160°C
Elastic modulus, E (GPa)	205.8	205.8	205.8	205.8
Yield stress, σ_{YC} (MPa)	359.6	382.0	387.2	411.5
Poisson's ratio (−)	0.3	0.3	0.3	0.3

Table 4.
Mechanical properties of AH32 steel at room and low temperatures in compression [7, 8].

remains unchanged regardless of sub-zero temperatures. This chapter focuses on the ultimate strength of steel structures under monotonically applied compressive loads, but fatigue crack resistance at sub-zero temperatures must be associated with microstructural characteristics which are closely related to low-temperature impact toughness of steel [52, 53].

In the present case study, an elastic-perfectly plastic material model was applied without considering the strain-hardening effect. To implement the material model, *MAT_PLASTICITY_COMPRESSION_TENSION, MAT124 in LS-DYNA was used as it is an isotropic elastic–plastic material which can distinguish material behaviour in tension and compression. The von Mises yield criterion was applied using MAT124. Tension or compression was determined by the sign of the mean stress (hydrostatic stress). A positive sign which means a negative pressure is indicative of tension, or a negative sign is indicative of compression. The mean stress, σ_{mean} can be expressed as follows:

$$\sigma_{mean} = \frac{\sigma_x + \sigma_y + \sigma_z}{3} \tag{16}$$

where σ_x, σ_y and σ_z are the stress components in the x, y and z directions, respectively.

Majzoobi et al. [41] observed that the ductile-brittle transition of steel occurs at −80°C, and the material behaviour of steel is completely brittle at −196°C. With **Figure 12** and **Table 2**, the average steel temperatures of plating and web of stiffeners in the middle bay of the tested structure were −160°C and −130°C, respectively. Therefore, the plating and web of stiffeners in the middle bay of the tested structure were modelled using the engineering stress-engineering stress relation of Eqs. (12) and (13). The rest of structural members in ductile region (above −80°C) were modelled using the engineering stress-engineering stress relation of Eqs. (12) and (13). See Paik et al. [7] for details.

Three types of fabrication-related initial deformations are considered as shown in **Figure 13**. The measurement data of welding-induced initial deformations for the tested structure [54] as shown in **Figure 14** was directly applied to the FE model.

The initial deformations of the tested structure were formulated so as to make easier implementation into the FE model as shown in **Figure 15**.

Figure 13.
Three types of welding-induced initial deformations in a stiffened plate structure.

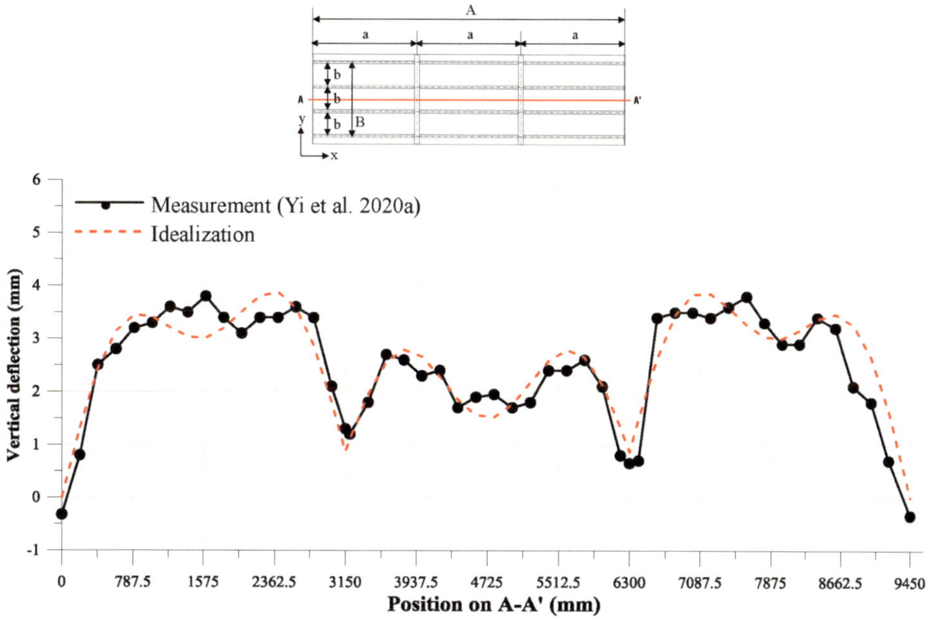

Figure 14.
Measured and idealised deformations of the tested structure due to fabrication by welding.

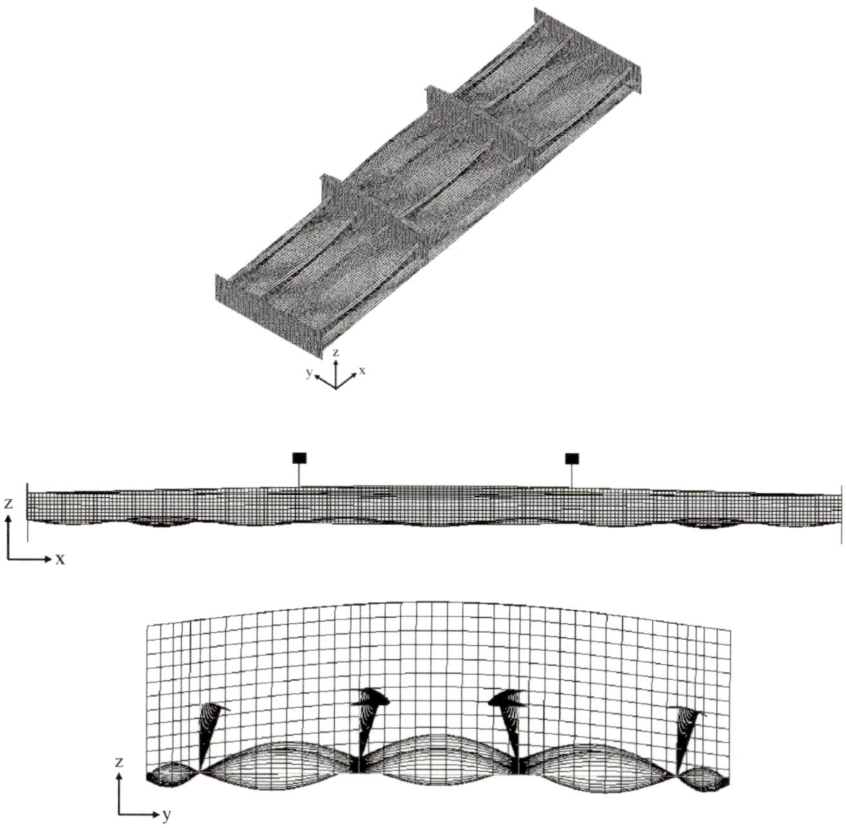

Figure 15.
Welding-induced initial deformations applied to the FE model (with an amplification factor of 100 for plating and column-type, and 20 for sideways initial deformations).

Initial deflection of plating:

$$w_{opl} = w_{opl}^{*} + w_{opl}^{**} \tag{17}$$

$$w_{opl}^{*} = w_{o\,\max} \sin\left(\frac{m\pi x}{a}\right) \left|\sin\left(\frac{n\pi y}{b}\right)\right| \text{ with } m = 1,\ n = 1,$$

$$w_{o\,\max} = \begin{cases} 3.5 \ \ for\ 0 \le x \le a, 2a \le x \le 3a \\ 1.5 \ \ for\ a \le x \le 2a \end{cases} \tag{18}$$

$$w_{opl}^{**} = w_{o\,\max} \sin\left(\frac{m\pi x}{a}\right) \left|\sin\left(\frac{n\pi y}{b}\right)\right| \text{ with } w_{o\,\max} = 1.0,\ m = 3,\ n = 1 \tag{19}$$

Column-type initial deformation of stiffener:

$$w_{oc} = w_{oc\,\max} \sin\left(\frac{m\pi x}{A}\right) \sin\left(\frac{n\pi y}{B}\right) \text{ with } w_{oc\,\max} = 1.0,\ m = 1,\ n = 1 \tag{20}$$

Sideways initial deformation of stiffener:

$$w_{os} = w_{os\,\max} \left(\frac{z}{h_w}\right) \sin\left(\frac{\pi x}{a}\right) \text{ with } w_{os\,\max} = 0.000133a \tag{21}$$

where z is the coordinate in the direction of stiffener web height, and h_w is the stiffener web height.

Biaxial residual stresses developed in the plating of the tested structure between the support members because the welding was conducted in both the longitudinal and the transverse directions to attach the longitudinal stiffeners and the transverse frames. Measurement data of the fabrication-induced residual stresses in the tested structure [55] was also directly applied to the FE model although the biaxial residual stress distributions were idealised as shown in **Figure 16** with the measurement data indicated in **Table 5**.

Stress concentration in structural details or fillet weld toe locations happens due to geometrical discontinuity, and it is a critical factor that must be considered for

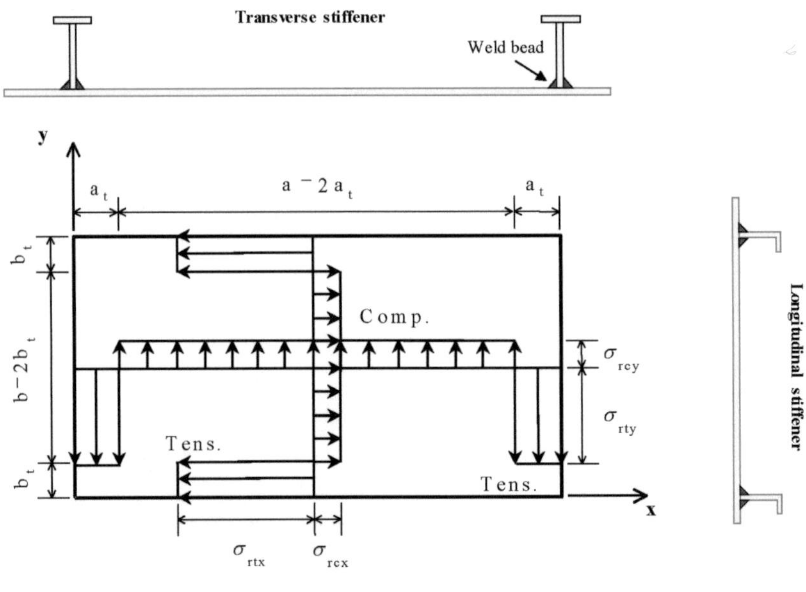

Figure 16.
Idealised distribution of biaxial residual stresses in plating of the tested structure.

	Transverse direction			Longitudinal direction	
	Simplified model	**Smith average level model**		**Simplified model**	**Smith average level model**
b_t	39.61 mm	56.80 mm	a_t	51.47 mm	64.70 mm
σ_{rcx}	$-0.110\ \sigma_{YT}$	$-0.150\ \sigma_{YT}$	σ_{rcy}	$-0.030\ \sigma_{YT}$	$-0.034\ \sigma_{YT}$
σ_{rtx}	$+0.90\ \sigma_{YT}$	$+0.80\ \sigma_{YT}$	σ_{rty}	$+0.90\ \sigma_{YT}$	$+0.80\ \sigma_{YT}$

Table 5.
Measured data of the biaxial residual stresses in the plating of the tested structure.

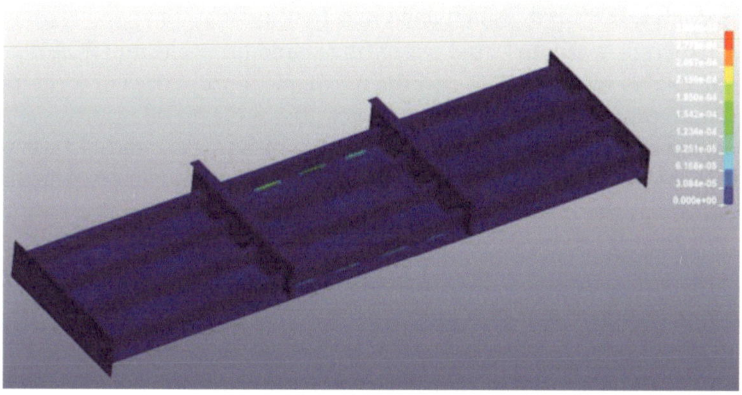

Figure 17.
Effective plastic strain (−) distribution in FE analysis on ultimate compressive strength of stiffened plate structure.

fatigue limit state analysis [3, 56]. **Figure 17** shows an example of the effective plastic strain distribution which was obtained from the FE analysis of the ultimate compressive strength of the tested structure. It is obvious from **Figure 17** that the effective plastic strain is comparatively large along the weld lines between plating and stiffeners. For the ultimate strength analysis in ductile region, e.g., at room temperature, the stress concentration at the fillet weld toes is usually ignored.

For brittle fracture analysis at sub-zero or cryogenic condition, however, the effects of stress concentration cannot be neglected [4, 44]. This is because the weld toes can reach the yield condition earlier, leading to local brittle fracture which can trigger the ultimate limit states at cryogenic condition. Therefore, the nonlinearity at weld toes along the fillet weld lines needs to take into account in the FE modelling.

One of approaches is to model the weld toes directly in the FE model using shell elements with specific properties of weld metal. **Figure 18** shows a schematic of

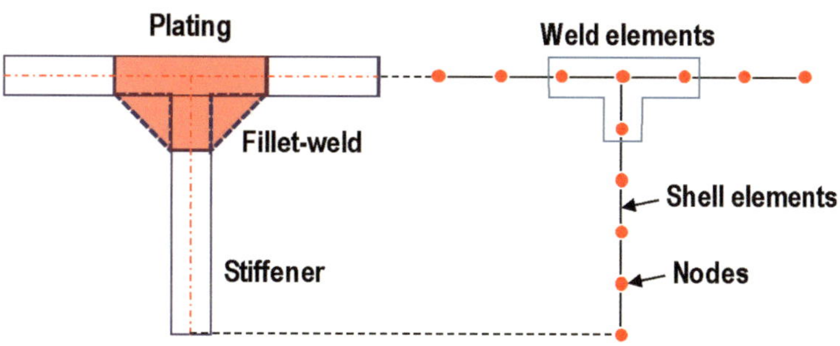

Figure 18.
Weld elements at the plate-stiffener junction.

	At 20°C	At −160°C (tension)	At −160°C (compression)
Elastic modulus, E (GPa)	268.0	268.0	268.0
Yield strength, σ_Y (MPa)	513.0	929.3	571.1
Poisson's ratio (−)	0.3	0.3	0.3

Table 6.
Mechanical properties of the weld metal with the CSF-71S at room temperature and assumed mechanical properties at −160°C.

modelling the weld toes using shell elements along the plate-stiffener junction. A similar approach was used to model weld toes by Kim et al. [44] and Nam et al. [35]. The tested structure was fabricated using flux-cored arc welding (FCAW) method and the consumable was CSF-71S, and the mechanical properties of weld metal with the CSF-71S are presented in **Table 6**. As such, the weld metal was modelled using the engineering stress-engineering strain relation of Eqs. (12) and (13). It is assumed that the yield strength of the weld metal at −160°C increases linearly in the same proportion as the steel (**Table 6**).

5. Computed results and discussion

Figure 19, **Tables** 7 and **8** present the comparison results between the test and the FE analysis. The difference of ultimate strength between them is 16.6% by the ductile material model, but it becomes at most 2.3% by the brittle material model.

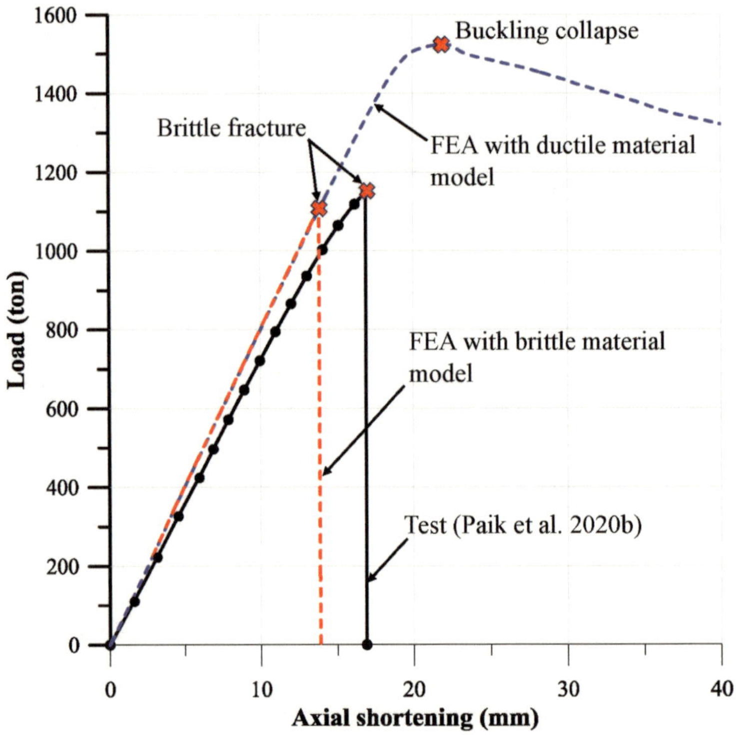

Figure 19.
Comparison of the load-axial shortening curves from the test and the FE analysis with the simplified brittle material model.

Parameter	Test [7]	FE analysis with ductile material model	Difference
Ultimate strength (ton)	1149.06	1340.14	+ 16.6%
Stiffness (ton/mm)	72.38	80.80	+ 11.6%
Axial shortening up to collapse (mm)	16.90	19.24	+ 13.8%
Strain energy up to collapse (ton·mm)	10179.96	14541.84	+ 42.8%
Local buckling	None	Occurs	—
Brittle fracture	Occurs	None	—

Table 7.
Comparison between the test and the FE analysis with the ductile material model.

Parameter	Test [7]	FE analysis with brittle material model	Difference
Ultimate strength (ton)	1149.06	1176.01	+ 2.3%
Stiffness (ton/mm)	72.38	80.82	+ 11.7%
Axial shortening up to collapse (mm)	16.90	14.75	- 12.7%
Strain energy up to collapse (ton·mm)	10179.96	8740.46	- 14.1%
Local buckling	None	None	—
Brittle fracture	Occurs	Occurs	—

Table 8.
Comparison between the test and the FE analysis with the brittle material model.

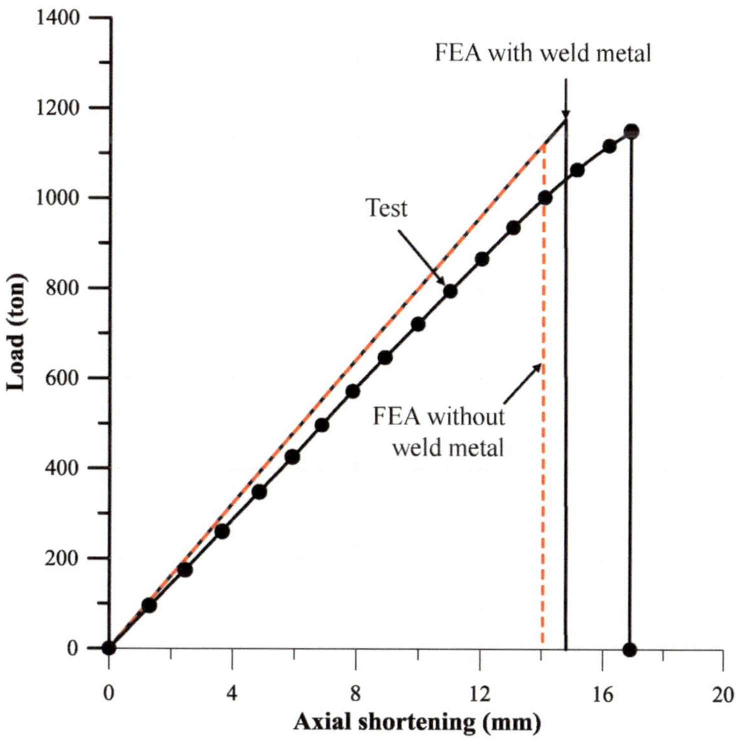

Figure 20.
Effect of weld metal on the ultimate strength behaviour.

Parameter	FEA with consideration of weld metal	FEA without consideration of weld metal	Difference
Ultimate strength (ton)	1176.01	1121.15	- 4.7%
Stiffness (ton/mm)	80.82	80.82	—
Axial shortening up to collapse (mm)	14.75	14.05	- 4.7%
Strain energy up to collapse (ton·mm)	8740.46	7908.81	- 9.5%
Local buckling	None	None	—
Brittle fracture	Occurs	Occurs	—

Table 9.
Effect of weld metal on the ultimate strength behaviour.

When only the ductile material model was applied for all structure members without considering brittle fracture, the FE analysis overestimates the ultimate strength significantly. As the yield strength of the material at cold temperature is greater than that at room temperature, the ultimate strength becomes much larger as far as brittle fracture is not allowed to happen. On the other hand, the ultimate strength obtained from the FE analysis with the brittle material model is in good agreement with the test results. **Figure 20** compares the ultimate strength behaviour with or without the weld elements along the plate-stiffener junctions. It is seen from **Figure 20** that the weld metal model increased the ultimate strength by 4.7%. This is due to the mechanical properties of weld metal which are larger than those of base metal (**Table 9**).

Figure 21 shows the deformed shape at the ultimate limit state of the tested structure obtained from the FE analysis without brittle fracture model (with only ductile material model). It is seen from **Figure 21** that the tested structure reached the ultimate limit state by tripping mode of stiffeners (without brittle fracture) which is similar to the collapse mode at room temperature [1]. However, the brittle fracture model represents brittle fracture behaviour which triggered the ultimate strength as shown in **Figure 22**, where deformed and fracture shapes of the test structures are compared between physical testing and FE analysis.

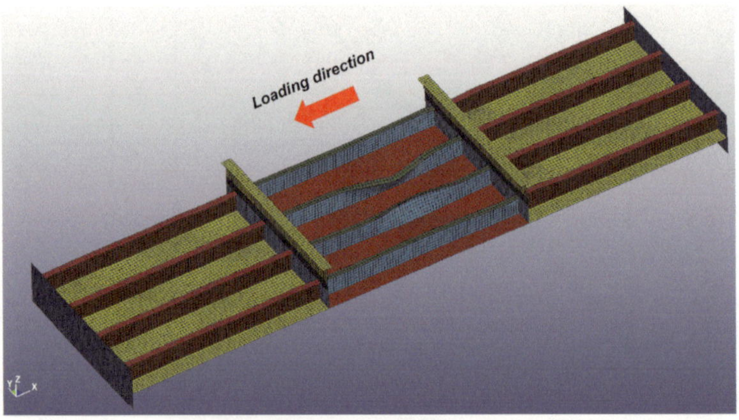

Figure 21.
Deformed shape of the tested structure at the ultimate limit state obtained from FE analysis only with ductile material model.

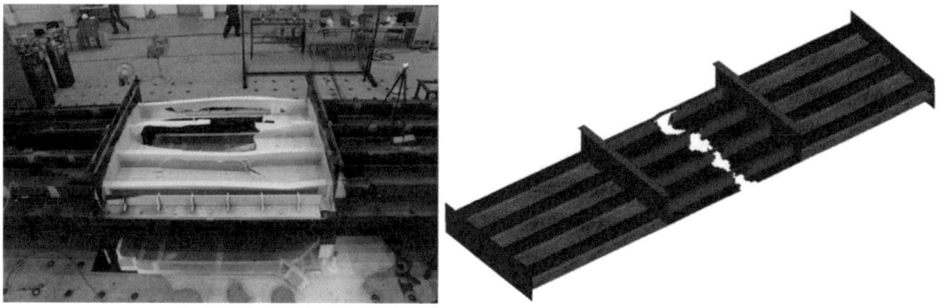

Figure 22.
Deformed shapes of the tested structure at the ultimate limit state obtained from the test and the FE analysis with brittle material model.

6. Concluding remarks

This chapter presents a practical method to compute the ultimate compressive strength of steel stiffened-plate structures at cryogenic condition which is triggered by brittle fracture. Case studies were carried out using the method, and the following conclusions were obtained together with modelling recommendations for NLFEM simulations.

1. A useful material model was formulated to analyse the brittle fracture behaviour of structural steels at cryogenic condition, where the Bauschinger effect was taken into account as the material properties in compression are distinct from those in tension

2. An elastic-perfectly plastic material model was applied without considering strain-hardening effect.

3. The developed material model was implemented into the LS-DYNA implicit code with *MAT_PLASTICITY_COMPRESSION_TENSION, MAT124.

4. Weld elements which are the same type of shell elements but with specific properties of material were introduced to model weld metal (weld toes) at the plate-stiffener junctions where stress concentrations develop. As the mechanical properties of weld metal are typically greater than those of base metal, the ultimate strength usually becomes larger with weld elements at weld toes.

5. Comparisons were made between the test results and the FE computations on a full-scale steel stiffened-plate structure at cryogenic condition. It is confirmed that the nonlinear FE analysis with the proposed material model, and the weld element model, gives a reasonably good solution of the ultimate compressive strength behaviour for steel stiffened plate structures at cryogenic condition.

Acknowledgements

This study was undertaken in the International Centre for Advanced Safety Studies, www.icass.center (the Korea Ship and Offshore Research Institute) at Pusan National University which has been a Lloyd's Register Foundation Research

Centre of Excellence since 2008. Part of the work was supported by the Swedish Research Council by the project "Fundamental research on the ultimate compressive strength of ship stiffened-plate structures at Arctic and cryogenic temperatures", contract no. 2018-06864.

Author details

Dong Hun Lee[1], Jeom Kee Paik[2,3*], Jonas W. Ringsberg[4] and P.J. Tan[2]

1 Generic Technology Research Center, Samsung Heavy Industries Co., Ltd., Geoje, South Korea

2 Department of Mechanical Engineering, University College London, London, UK

3 The Korea Ship and Offshore Research Institute (Lloyd's Register Foundation Research Centre of Excellence), Pusan National University, Busan, South Korea

4 Department of Mechanics and Maritime Sciences, Chalmers University of Technology, Gothenburg, Sweden

*Address all correspondence to: j.paik@ucl.ac.uk

References

[1] Paik JK, Lee DH, Noh SH, Park DK, Ringsberg JW. 2020a. Full-scale collapse testing of a steel stiffened plate structure under cyclic axial-compressive loading. Structures, 26: 996–1009.

[2] Hughes OF, Paik JK. 2013. Ship structural analysis and design. Alexandria, USA: The Society of Naval Architects and Marine Engineers.

[3] Paik JK. 2018. Ultimate limit state analysis and design of plated structures, 2nd Edition. Chichester, UK: John Wiley & Sons.

[4] Paik JK. 2020. Advanced structural safety studies: with extreme conditions and accidents. Singapore: Springer.

[5] Paik JK. 2021. Ship-shaped offshore installations: design, construction, operation, healthcare and decommissioning. Cambridge University Press, Cambridge, UK.

[6] Smith CS, Davidson PC, Chapman JC, Dowling PJ. 1988. Strength and stiffness of ships' plating under in-plane compression and tension. RINA Transactions. 130: 277–296.

[7] Paik JK, Lee DH, SH, Noh, DK Park, Ringsberg JW. 2020b. Full-scale collapse testing of a steel stiffened plate structure under axial-compressive loading triggered by brittle fracture at cryogenic condition. Ships and Offshore Structures, 15(S1): S29-S45.

[8] Paik JK, Lee DH, Park DK, Ringsberg JW. 2020c. Full-scale collapse testing of a steel stiffened plate structure under axial-compressive loading at a temperature of -80°C. Ships and Offshore Structures, doi: 10.1080/17445302.2020.1791685.

[9] DNVGL (2017). Rules for classification – ships. Part 2 Materials and welding, Chapter 4 Fabrication and testing. Oslo, Norway.

[10] Brokenbrough RL, Johnston BG. 1981. USS steel design manual. United States Steel Corporation, Pittsburgh, PA, USA.

[11] Dowling NE. 2012. Mechanical behavior of materials, 4th Edition. London, UK: Pearson.

[12] Cerik BC, Choung JM. 2020. On the prediction of ductile fracture in ship structures with shell elements at low temperatures. Thin-Walled Structures. 151. doi: 10.1016/j.tws.2020.106721.

[13] Chen L, Liu W, Yu L, Cheng Y, Ren K, Sui H, Yi X, Duan H. 2020. Probabilistic and constitutive models for ductile-to-brittle transition in steels: A competition between cleavage and ductile fracture. Journal of the Mechanics and Physics of Solids. 135: 103809.

[14] Choung JM, Nam W, Lee JY. 2013. Dynamic hardening behaviors of various marine structural steels considering dependencies on strain rate and temperature. Marine Structures. 32: 49–67.

[15] Yoo SW, Lee CS, Park WS, Kim MH, Lee JM. 2011. Temperature and strain rate dependent constitutive model of TRIP steels for low-temperature applications. Computational Material Science. 50(7): 2014–2027.

[16] Ehlers S, Varsta P. 2009. Strain and stress relation for non-linear finite element simulations. Thin-Walled Structures, 47: 1203–1217.

[17] Ehlers S. 2010. Strain and stress relation until fracture for finite element simulations of a thin circular plate. Thin-Walled Structures, 48: 1–8.

[18] Bai Y, Wierzbicki T. 2008. A new model of metal plasticity and fracture

with pressure and Lode dependence. International Journal of Plasticity, 24 (6): 1071–1096.

[19] Bai Y, Wierzbicki T. 2010 Application of extended Mohr–Coulomb criterion to ductile fracture. International Journal of Fracture, 161: 1–20.

[20] Coppola T, Cortese L, Folgrarait P. 2009. The effect of stress invariants on ductile fracture limit in steels. Engineering Fracture Mechanics, 76(9): 1288–1302.

[21] Gruben G, Hopperstad OS, Børvik T. 2012. Evaluation of uncoupled ductile fracture criteria for the dual-phase steel Docol 600DL. International Journal of Mechanical Sciences, 62(1): 133–146.

[22] Lou Y, Huh H, Lim S, Pack K. 2012. New ductile fracture criterion for prediction of fracture forming limit diagrams of sheet metals. International Journal of Solids and Structures, 49(25): 3605–3615.

[23] Lou Y, Huh H. 2013. Extension of a shear-controlled ductile fracture model considering the stress triaxiality and the Lode parameter. International Journal of Solids and Structures, 50(2): 447–455.

[24] Voyiadjis GZ, Hoseini SH, Farrahi GH. 2012. Effects of stress invariants and reverse loading on ductile fracture initiation. International Journal of Solids and Structures, 49(13): 1541–1556.

[25] Cerik BC, Lee KS, Park SJ, Choung JM. 2019a. Simulation of ship collision and grounding damage using Hosford-Coulomb fracture model for shell elements. Ocean Engineering, 173: 415–432.

[26] Cerik BC, Park BJ, Park SJ, Choung JM. 2019b. Modeling, testing and calibration of ductile crack formation in grade DH36 ship plates. Marine Structures, 66: 27–43.

[27] Cerik BC, Ringsberg JW, Choung JM. 2019c. Revisiting MARSTRUCT benchmark study on side-shell collision with a combined localized necking and stress-state dependent ductile fracture. Ocean Engineering, 187. doi: 10.1016/j.oceaneng.2019.106173.

[28] Cerik BC, Park SJ, Choung JM. 2020. Use of localized necking and fracture as a failure criterion in ship collision analysis. Marine Structures, 73: doi: 10.1016/j.marstruc.2020.102787.

[29] Choung JM. 2009. Comparative studies of fracture models for marine structural steels. Ocean Engineering, 36: 1164–1174.

[30] Choung JM, Shim CS, Song HC. 2012. Estimation of failure strain of EH36 high strength marine structural steel using average stress triaxiality. Marine Structures, 29: 1–21.

[31] Noh MH, Cerik BC, Han DH, Choung JM. 2018. Lateral impact tests on FH32 grade steel stiffened plates at room and sub-zero temperatures. International Journal of Impact Engineering, 115: 36–47.

[32] Park SJ, Cerik BC, Choung JM. 2020. Comparative study on ductile fracture prediction of high-tensile strength marine structural steels. Ships and Offshore Structures, doi: 10.1080/17445302.2020.1743552.

[33] Kaminskji AA, Galatenko GV. 1999. On the temperature dependence fracture toughness in the brittle-to-ductile transition region. International Applied Mechanics. 35(4): 398–404.

[34] KSNA. 1983. Handbook for ship design. 4th Edition, The Kansai Society of Naval Architects, Osaka, Japan (in Japanese).

[35] Nam W, Hopperstad OS, Amdahl J. 2018. Modelling of the ductile-brittle fracture transition in steel structures

with large shell elements: A numerical study. Marine Structures. 62: 40–59.

[36] Nazari A, Milani AA. 2011. Ductile to brittle transition temperature of functionally graded steels with crack arrester configuration. Materials Science and Engineering: A. 528(10–11): 3854–3859.

[37] Perez-Martin MJ, Holmen JK, Thomesen S, Hopperstad OS, Børvik T. 2019. Dynamic behaviour of a high-strength structural steel at low temperatures. Journal of Dynamic Behavior of Materials. 5: 241–250.

[38] Tanguy B, Besson J, Piques R, Pineau A. 2005a. Ductile to brittle transition of an A508 steel characterized by Charpy impact test: Part I: experimental results. Engineering Fracture Mechanics. 72(1): 49–72.

[39] Tanguy B, Besson J, Piques R, Pineau A. 2005b. Ductile to brittle transition of an A508 steel characterized by Charpy impact test: Part II: modeling of the Charpy transition curve. Engineering Fracture Mechanics. 72(3): 413–434.

[40] Tong L, Niu L, Jing S, Ai L, Zhao, XL. 2018. Low temperature impact toughness of high strength structural steel. Thin-Walled Structures. 132: 410–420.

[41] Majzoobi GH, Mahmoudi AH, Moradi S. 2016. Ductile to brittle failure transition of HSLA-100 steel at high strain rates and subzero temperatures. Engineering Fracture Mechanics. 158: 179–193.

[42] Paik JK, Kim BJ, Park DK, Jang BS. 2011. On quasi-static crushing of thin-walled steel structures in cold temperature: experiment and numerical studies. International Journal of Impact Engineering, 38: 13–28.

[43] Park DK, Kim DK, Park CH, Jang BS, Kim BJ, Paik JK. 2015. On the crashworthiness of steel-plated structures in an Arctic environment: an experimental and numerical study. Journal of Offshore Mechanics and Arctic Engineering, 137: 051501–1–051501-11.

[44] Kim KJ, Lee JH, Park DK, Jung BK, Han X, Paik JK. 2016. An experimental and numerical study on nonlinear impact responses of steel-plated structures in an Arctic environment. International Journal of Impact Engineering. 93: 99–115.

[45] Juvinal RC, Marshek K. 2011. Fundamentals of machine component design, 5th Edition. Chichester, UK: John Wiley & Sons.

[46] Johnson GR, Holmquist TJ. 1994. An improved computational constitutive model for brittle materials. High-Pressure Science and Technology, American Institute of Physics.

[47] Deshpande, V. S., Evans, A. G. (2008). 'Inelastic deformation and energy dissipation in ceramics: A mechanism-based constitutive model'. Journal of Mechanics and Physics of Solids, 56(10): 3077–3100.

[48] Bhat HS, Rosakis AJ, Sammis CG. (2012). 'A micromechanics based constitutive model for brittle failure at high strain rates'. Journal of Applied Mechanics, 79(3): 031016, doi: 10.1115/1.4005897.

[49] Paik JK, Kim KJ, Lee JH, Jung BG, Kim SJ. 2017. Test database of the mechanical properties of mild, high-tensile and stainless steel and aluminium alloy associated with cold temperatures and strain rates. Ships and Offshore Structures, 12(S1): S230-S256.

[50] Ehlers S, Østby E. 2012. Increased crashworthiness due to arctic conditions–The influence of sub-zero temperature. Marine Structures. 28: 86–100.

[51] von Mises R. 1913. Mechanik der festen Körper im plastisch-deformablen Zustand. Nachrichten von der Gesellschaft der Wissenschaften zu Göttingen. Mathematisch-Physikalische Klasse. (1): 582–592.

[52] Jorge JCF, de Souza LFG, Mendes MC, Bott IS, Araújo LS, dos Santos VR, Rebello JMA, Evans GM. 2021. Microstructure characterization and its relationship with impact toughness of C-Mn and high strength low alloy steel weld metals – a review. Journal of Materials Research and Technology. 10: 471–501.

[53] Kang YJ, Park GT, Jeong SH, Lee CH. 2018. Correlation between microstructure and low-temperature impact toughness of simulated reheated zones in the multi-pass weld metal of high-strength steel. Metallurgical and Materials Transactions A, 59: 177–186.

[54] Yi MS, Lee DH, Lee HH, Paik JK. 2020. Direct measurements and numerical predictions of welding-induced initial deformations in a full-scale steel stiffened plate structure. Thin-Walled Structures, doi: 10.1016/j.tws.2020.106786.

[55] Yi MS, Noh SH, Lee DH, Seo DH, Paik JK. 2021. Direct measurements and numerical predictions and simple formula estimations of welding-induced biaxial residual stresses in a full-scale steel stiffened plate structure. Structures, 29: 2094–2105.

[56] Lotsberg I. 2016. Fatigue design of marine structures. Cambridge University Press, Cambridge, UK.

Section 2

Low-Temperature Refrigeration and Applications

Chapter 5

Refrigerant Mixtures

Selva Pandian Ebenezer

Abstract

Evolution of refrigerants has a history since the introduction of air conditioning systems by Sir Willis Carrier. The first generation air conditioning systems used natural refrigerants like air, water, carbon dioxide etc. But the need of low temperature requirements in residential and industrial air conditioning systems has forced the air conditioning field to use chloro flouro carbon type refrigerants, which was introduced by Dupon in the previous century. Physical and chemical properties of CFC type refrigerants were very good and satisfactory and so it was used in almost all refrigeration and air conditioning systems. The main disadvantage of CFC type refrigerants is harming the ozone layer and contributing much to the global warming. This chapter reviews the use of CFC type and introducing alternate type of refrigerants.

Keywords: Refrigerants, Chloro Flouro Carbon, Ozone, Global Warming, Emission, Environment, Alternate refrigerants

1. Introduction

Carbon dioxide (CO_2) is a naturally arising gas by the method of photosynthesis into organic matter. A derivative of fossil fuel ignition and biomass burning, it is also released from the changes in the use of lands and other industrial activities. Earth's radiative stability is continuously disturbed primarily by carbon di oxide. CO_2 is considered to be a reference gas for the measurement of other greenhouse gases and thus having a Global Warming Potential of 1. The rate of global warming increase is because of the climate change and escalation in the concentrations of atmospheric carbon dioxide.

This is because of the increase in the custom of using carbon based fuels especially in the present modern world. CO_2 is also a key cause of marine acidification as it softens in water to produce carbonic acid. The earth's radiative balance gets disturbed because of the continuous addition of greenhouse gases in the atmosphere. As a result, we observe an increase in the earth's surface temperature and extreme changes in climate, rise in sea levels, and harmful effects on world agriculture. Since the past two decades, global emissions of carbon dioxide have risen by 99%, or on an average 2.0% for a year, and it is expected to rise by another 45% by the end of 2030, or increase in the rate of 1.6% per year.

2. History and evolution of chloro flouro carbons (cfc)

The emergence of chloro flouro carbons which are popularly known as CFC has not happened in a day. Natural refrigerants like water, carbon di oxide were used in

Sl.No	Year	Refrigerant	Comment
1	1930	—	Announcement of the development of Fluorocarbon refrigerant
2	1931	R-12	Commercial refrigerant
3	1932	R-11	Commercial refrigerant
4	1933	R-114	Commercial refrigerant
5	1934	R-113	Commercial refrigerant
6	1936	R-22	Commercial refrigerant
7	1943	R-11 & R-12	Developed to use as an aerosol propellants
8	1945	R-13	Introduced as a commercial refrigerant
9	1949	R-500	Patented by carrier corporation
10	1952	—	Manufacture of fluorocarbon refrigerants
11	1955	R-14	Introduced as a commercial refrigerant
12	1957- 1963	—	Production of fluorocarbon refrigerants started by other industries
13	1961	R-502	Introduction of R-502 in a commercial manner
14	1975	R-12 & R-13	Thermodynamic properties were established

Table 1.
History and evolution of chloro flouro carbons.

refrigeration and air conditioning industry in the past which did not impose any destructive effects to the environment and ozone layer. Later on, chloro flouro carbons were introduced by the company Dupon in the year 1930, as a result of remarkable developments in the refrigeration industry. CFC's were released to the atmosphere carelessly during service and repairing of refrigeration and air conditioning equipment. Then it reaches the topmost layer of the atmosphere and destroys the ozone layer, which leads to many harmful effects to humans, animals and crop cultivation. Below **Table 1** shows the brief history and the evolution of chloro flouro carbons.

3. Need of moving towards natural refrigerants

In early 1970s, scientists come to know about the hazardous impacts contributed to the earth's atmosphere by Chlorofluorocarbons (CFCs). CFCs were widely used as foams, refrigerants and thinners for many industrial applications. UV-B radiation which is passing through the ozone depletion areas from the sun can spread straight to the Earth's surface and will cause distinctive harms in the human cells, plants and animals. And this is because of the ozone layer destruction by CFC's. An international treaty was decided at Montreal, Canada, to ensure further damage of ozone layer. The outcome of Montreal protocol was to begin the phasing-out the usage of CFCs and other Ozone Depleting Substances (ODS) like Hydro chloro fluorocarbons (HCFCs).

Hydro fluoro carbons (HFCs) are considered to be one of the major, fastest growing, and most potent, greenhouse gases. In the past two decades, discharges of hydro fluoro carbons (HFCs) have been increasing swiftly. HFC's are the substitutes for chlorofluorocarbons and hydro chloro fluoro carbons (HCFCs). But stratospheric ozone is not destroyed by HFC's, but they are considered to be one of the effective greenhouse gases with a significant global warming potential (GWP) [1]. Many commercial refrigeration systems, such as beverage coolers, vending machines, ice cream freezers, open deck coolers and freezers used in hypermarkets

Refrigerant	ODP (ozone depletion potential)	GWP (global warming potential)
CFC	High	Very High
HCFC	Very Low	Very High
HFC	Zero	High
HC	Zero	Insignificant
CO_2	Zero	Insignificant
Helium used in Coolers	Zero	Zero

Table 2.
ODP and GWP of popular refrigerants.

use Hydro chloro fluorocarbons (HCFCs) and Hydro fluoro carbons (HFCs) gases as refrigerants. HCFCs are one of the ozone depleting agents and they have to be phased out as per the Montreal Protocol.

Even though the phasing out of CFC's are a great success in developed countries, there is still a delay in phasing out of HCFC's in the developing countries. There is a need of awareness among the general public about the safe use of fluorinated refrigerants and their adverse effects to our planet earth. Necessary training for the refrigeration and air conditioning technicians in the developing countries as this will ensure the safe recovery of these harmful refrigerants. There is a repeated history prevailing now in the use of natural refrigerants in the air conditioning industry. Because of the climatic changes and associated global warming problems, nowadays natural refrigerants are preferred in place of CFC's. In the present technology in the refrigerant sector, two ozone-friendly refrigerant technologies are available instead of CFC's: 1. Fluorinated refrigerants (HFC's) which are harmful to climate and 2. Natural refrigerants which not harmful to environment. So it is obvious and mandatory to move towards the use of natural refrigerants which have advantages on climate ozone layer. The **Table 2** as shown below present the ODP and GWP of popular refrigerants.

4. Green house gas emission

Green house gas emission is because of six potential gases. Green house gas emission in Oman during 2000's was around 30 million metric tons [2]. Refrigerant leakage can be a small quantity, but it can be a considerable source of greenhouse

Symbol	Name	Common Sources
CO_2	Carbon Dioxide	Resultant gas from combustion, manufacture of cement products, Etc.
CH_4	Methane	Landfills, manufacture and refining of natural gas and petroleum, fermentation from the digestive system of livestock, cultivation of rice, resultant gas from combustion, etc.
N_2O	Nitrous Oxide	Gas output from combustion, fertilizers, manufacture of nylon, manure, etc.
HFC's	Hydro fluoro carbons	Refrigerants, smelting of aluminum, manufacturing of semiconductor devices, etc.
PFC's	Fluorocarbons	Aluminum production, semiconductor industry, etc.
SF_6	Sulfur Hexafluoride	Transmissions and distribution of electrical systems, circuit breakers, production of magnesium etc.

Table 3.
Sources and properties of greenhouse gases.

gas emission. The below **Table 3** show the common sources and their properties which are responsible for the greenhouse gas emission.

5. Global warming potential

Carbon di oxide is considered to be the knob of earth's thermostat and it is an amazing tracer gas. Even a small change in the Co_2 concentrations makes a big difference to the global surface temperature [3]. Greenhouse gases have the properties of active radiative or heat-trapping nature. Comparing the properties of greenhouse gases, are done by indexing them according to their Global Warming Potential. The ability of a GHG to trap heat in the atmosphere comparative to an equivalent quantity of carbon dioxide is called GWP. Carbon dioxide has the value one (1), though the most prevalent, is the least powerful GHG. So, the greenhouse gases are expressed in carbon dioxide equivalents. The unit of GWP potential is million metric tons (MMTCDE) of carbon dioxide and greenhouse gas emission from an electrical appliance can be calculated by using the formula,

$$\frac{\text{Hours} \times \text{No.of days} \times \text{Watts}}{1000} \text{KWh}$$

$$\text{KWh} \times \text{Emission factor} = \text{Kg of } Co_2$$

Earth's life is protected from sun's harmful UV rays by ozone layer which is formed as a thin layer in the stratosphere. Ozone layer depletion was identified by the scientists during 1980. As a result of this, depletion of earth is likely to receive more amount of UV radiation, so that there is a strengthened chance of overexposure to UV radiation and the subsequent wellbeing effects. The below **Figure 1** depicts the formation and destruction of ozone depletion process. The sun's yield of UV B does not change. It is obvious that, less ozone means, more exposure of UV B radiation from the sun. The amount of UV B measured at the surface of Antarctic poles is two times during the annual ozone hole.

5.1 Mechanism of ozone depletion by CFC's

Ozone layer is destroyed in the stratosphere 15 to 20 Km directly above the earth surface by CFC's. Ozone concentrations are measured in Dobson units. 1 Dobson unit denotes 1 ozone molecule for every 1 billion air molecules. The meaning of ozone hole is the loss of ozone in a particular area. Greatest ozone hole is recorded in Antarctica continuously. The characteristics of ozone are, it is an allotrope of oxygen, and it is deadly to human beings if it is inhaled [4]. Human beings existence in the earth is very important and this is ensured by the protective layer ozone, as it filters or captivates ultra violet radiations which are usually short in wave length (280 – 320 nm). Ultra violet radiations can cause serious problems to humans such as sun burns, skin cancer, and eye disorders.

Figure 1.
Ozone formation and depletion process.

One of the chief characteristics of Chloro flouro carbons is, they do not dissolve in water and highly inert to water solubility. That is the reason they are not destroyed or dissolved even during in rain and stay in the atmosphere for many years and move slowly towards the stratosphere. The chloro flouro carbon molecules split off into chlorine atoms from the CFC molecules when they come in to contact with the ultra violet rays. The primary split of the CFC molecules are shown in the below equation.

$$CCL_3F(g) \xrightarrow{\text{UV radiation}} CCl_2F\ (g) + Cl_{(g)}$$

Ozone layer is destroyed particularly by these single chlorine molecules.

$$Cl\ (g) + O_3(g) \rightarrow ClO + O_2(g)$$

Considerable amount of oxygen atoms are present in the stratosphere, because it produces oxygen atoms regularly by go through photo chemical breakdown. This will lead to the renewal of chlorine atoms in the stratosphere. So a lone CFC molecule can dismiss many ozone molecules.

$$ClO\ (g) + O \rightarrow O_2 + Cl\ (g)$$

5.2 Photochemistry of ozone and Chapman's cycle for ozone loss

Ozone was first discovered by a German chemist, Christian Friedrich in the year 1939. Ozone present in the stratosphere is formed by the chemical reactions between oxygen and sunlight. The production of stratospheric ozone is because of the chemical reaction balance.

One ozone molecule is broken into one oxygen molecule and the remaining oxygen molecule is absorbed by the ultra violet radiation [5]. The photochemistry of ozone depends on the interaction of sun's radiation with the in atmospheric gases, particularly with oxygen.

A clear understanding of ozone layer was first assumed by Chapman in the year 1930. According to Chapman, when oxygen is hit by high energy photon, it is fragmented into two oxygen atoms. The below equation depicts Chapman's reaction.

$$O_2 + \frac{hc}{\lambda} = O + O$$

Where h = plank's constant.
C = Speed of light (wavelengths shorter than 242 nano meters)

$$\lambda = \text{wavelength of photon}$$

Ozone production by solar ultra violet radiation produce more amounts of ozone than the actual amount of ozone present in the atmosphere. The production of ozone is balanced by ozone loss.

$$O3 + O \rightarrow O2 + O2$$

5.3 Effects of ozone layer weakening

5.4 Effects on human health

UVB causes nonmelanoma skin cancer and plays a major role in malignant melanoma development and it is evident by many Laboratory and epidemiological studies. Also, UVB causes eye cataracts. In general, sunlight contains some UVB, even with normal ozone levels. And it is advisable that there should be always a limit to the exposure to sun.

5.5 Effects on plants and marine creatures

Even a small amount of UV B radiation present in the sunlight will cause an impact in the physiological and evolving processes of plants. Damages to early developmental stages of fish, shrimp, crab, amphibians and other animals are caused by UV B rays. Some of the most severe effects of UV B radiation exposure to plants and animals are decrease in their reproduction capacity and reduced larval development. Even a small increase in UV (B) exposure could result in significant reduction in the size of the population of animals that eat these minor creatures.

5.6 Hydrocarbons as refrigerants

CFC -12 is having high ODP and very high GWP. Promising substitutes for CFC-12 are hydro carbons which do not have any halogen compounds. The satisfactory characteristics of hydrocarbons are environmentally safe, energy efficient, technologically reliable refrigerants. Hydrocarbons arise naturally formed from solidified plant matter, and throughout the world initiate as oil and natural gas. Flammability is the characteristic feature of HC type refrigerants and it is the concerning point to be considered as a refrigerant in air conditioning and refrigeration systems, even though it exhibits very low GWP values [6].

Lower paraffin's such as propane, butane, and isobutene were successfully used as refrigerant before the arrival of CFCs. The thermodynamic properties of the hydrocarbons are much better than any of the other alternatives to CFC's. Hydrocarbons are available at low price all over the world and are compatible with commonly used lubricants and materials of construction used in refrigeration systems. Hydrocarbons are very economical and they are readily available in most parts of the world.

6. Phase change materials as refrigerants

Using phase change materials in air conditioning systems can be an effective method for improving the process of cooling as well as minimizing the size of the system. The nature of phase change materials is they will absorb, store and release large amount of heat. Generally, the temperature of PCM increases with the increase in the ambient temperature. Researchers conducted various studies in using phase change materials for the use in air conditioning systems. Co_2 emission and power consumption of the systems using phase change materials are better in comparison with the conventional systems [7]. Around 7% of electrical power consumption reduction was observed in the air conditioner designed by Nataohorn Chaiyat and Tanongkiat Kiatsiriroat with PCM bed in comparison with the normal air conditioner [8]. During the transition period, the PCM melts and so absorbs heat. A reverse process happens when the PCM temperature is decreased [7]. Thermodynamic properties of CFC and Hydrocarbons are mentioned in the below **Table 4**.

6.1 Applications of PCM

1. **PCM used in wall systems:** PCM's have been successfully used in the middle layer of the wall systems and it is evident from many studies. PCM embedded in the wall systems can reduce the indoor air temperature up to 4.2°C [9].

Refrigerant	Critical temp (°C)	Boiling temp (°C)	Density (kg/m^3)	Heat of vaporization (kj/kg)
Propylene	91.4	−47.8	1.955	440.16
Propane	96.8	−42.1	2.019	425.92
I-Butylene	146.6	−6.3	2.550	391.58
Isobutylene	144.7	−7.0	2.500	397.02
Isobutene	135.0	−11.7	2.668	366.03
n-Butane	152.0	−0.5	2.703	387.81
R-12	112.0	−29.7	6.240	166.0

Table 4.
Thermodynamic properties of CFC and hydro carbons.

2. **Better energy performance in air conditioners:** Energy reduction in air conditioners can be achieved by using paraffin wax as a phase change material. A previous study reveals that the PCM temperature and the temperature of the air leaving the PCM bed were satisfactory around 2.73% and 4.61%, respectively. The cost of electricity saving of the improved system was about 9.10% when compared with the standard system [10].

3. **Use of phase change materials in refrigerators:** A refrigerator was designed and developed by Azzouz et al. and it was observed that, the compressor running time was decreased about 25% when compared with the conventional systems.

4. **Latent heat thermal storage:** Phase change materials are used widely for the latent heat storage systems in heat pumps, solar energy systems etc.

5. **Medical Industry:** In medical field, the energy storage characteristics of PCM is used for transporting blood, and hot & cold therapies [11]

6. **PCM used for energy efficient housing:** PCM's are widely used for energy efficient housing applications. Thermo-chromic PCM's are used as window coatings for better visual performance. Solid to liquid PCM's are used for the latent thermal energy storage and constant temperature applications [12].

7. Application of hydrocarbons as a refrigerant in commercial refrigeration systems

Worldwide attempts are being made to eliminate the use of Chlorofluorocarbons (CFCs) because chlorine released from CFCs migrates to the stratosphere and destroys the stratospheric ozone layer. An international treaty known as Montreal Protocol was formed to regulate the production and trade of ozone-depleting substances. Sultanate of Oman is one among the signatories of the Montreal Protocol. During July 2003 a workshop was organized by Ministry of Regional Municipalities, Environment & Water Resources in collaboration with the UNIDO and UNEP at Muscat to train the trainers to phase out the CFCs in Sultanate of Oman. The author was also one of the participants of the workshop. After the workshop it is felt to review the available literature of the various refrigerants, especially hydrocarbons, as alternative to CFCs and hence this article. This article is primarily intended to provide a brief summary about completed /going on works during the last toe decades to use hydrocarbon as refrigerants in commercial refrigeration systems.

Among the commonly used CFCs, **dichlorodifluoromethane** (CFC-12) is one of the most widely used refrigerants in various applications such as Domestic Refrigerators, Bottle, Coolers, Deep Freezers, Water Coolers, and Mobile Air Conditioners etc. The excellent characteristics of CFC – 12 have lead to the development of highly efficient and reliable compressors and other refrigeration system components. Studies have shown that these refrigeration appliances give satisfactory performances for approximately 15 to 20 years. This high degree of reliability has caused the consumers to expect long services from these appliances in general. This necessitates extensive evaluation of alternative refrigerants before adopting them for commercial use. **The ideal substitute for CFC-12 should be non-toxic, non-flammable, chemically stable, compatible with refrigeration system materials and lubricants and have transport and thermodynamic properties similar to or better than CFC-12.** In addition, the ideal substitute should have zero Ozone Depleting Potential (**ODP**) and low Global Warming Potential (**GWP**). However, there is no such single substance, which possesses all these properties.

Though, the CFCs were characterized in 1890, but the development of fluorocarbon refrigerants was announced in 1930. Since then CFCs never looked back. Some of the historical highlights in the progress of refrigeration and the development of refrigerants are outlined in **Table 5**.

The prevalent refrigerants and refrigerant mixtures from halogenated hydrocarbon family in use are R-11, R-12, R-13, R-14, R-22, R-113, R-114, R-500 and R-502. In developed countries various steps have been already taken to control the use of ozone depleting refrigerants. In developing countries conversion from CFCs to alternatives is still a major issue.

S.#	Year	Refrigerant	Comment.
1	1930		The development of Fluorocarbon refrigerant was announced.
2	1931	R-12	Introduced as a commercial refrigerant.
3	1932	R-11	Introduced as a commercial refrigerant.
4	1933	R-114	Introduced as a commercial refrigerant.
5	1934	R-113	Introduced as a commercial refrigerant.
6	1936	R-22	Introduced as a commercial refrigerant.
7	1943	Mixture of R-11 & R-12	Developed to use as an aerosol propellants
8	1945	R-13	Introduced as a commercial refrigerant.
9	1949	R-500	Patented by carrier corporation
10	1952	—	Manufacture of fluorocarbon refrigerants started by Allied Chemical Corporation.
11	1955	R-14	Introduced as a commercial refrigerant.
12	1957,1958, 1963		Manufacture of fluorocarbon refrigerants started by other companies.
13	1961	R-502	Commercial Introduction of R-502
14	1975	Mixture of R-12 & R-13	Establishment of thermodynamic properties over the whole range of composition.

Table 5.
Historical development of refrigerants.

As per the recent guidelines from the Environment Protection Agency USA, phasing down of HFCs and manufacturing of alternate refrigerants to CFCs and HFCs is the most significant environment policy to be practiced globally [13].

Montreal Protocol asks for abandoning the use and production of ODS in phased manner. It currently has the following control schedules for chemicals used as refrigerants:

- a phase out by 1.1.1996 of CFCs in the developed countries.
- a grace period until 2010 for a CFC phase out in the countries operating under paragraph 1 of Article 5 (the developing countries), with a freeze in 1999 and gradual reduction steps thereafter.
- a HCFC control schedule for the developed countries which requires gradual phase out of HCFCs over the period 1996 – 2020 (a freeze in 1996, a virtual phase out by 2020, a complete phase out by the year 2030), based upon a cap of 2.8% of the 1989 CFC consumption and the 1989 HCFC consumption (in ODP- tones);
- a HCFC control schedule for the developing countries, which lags that of the developed countries by 10 years.

The Montreal Protocol does not address non-ozone depleting chemicals According to Mc Linden over thirty years of research and development will be required to arrive at and maintain the family of refrigerants which are being used today.

Global action plans and the action plans to reduce the use and step by step phasing out of HFC type refrigernats was agreed by all the countries under the 2015, Paris amendment. In the year 2016, in Kigali, Ruwanda, around 197 countries had accepted for an amendment for the phasing down of HFC type refrigerant which was recommended by the Montreal Protocol. Phasing down of HFC type refrigerants has to be completely executed within the next 30 years [13].

8. HFC 134A: Present day refrigerant

HFC-134a (Tetra fluro ethane) presently is the leading candidate to replace CFC-12. The main culprit chlorine atom is absent in the molecule of HFC-134a, hence this substance provides excellent chemical and thermal stability, significantly better than CFC-12.It has got zero Ozone Depleting Potential (ODP) and Global Warming Potential (GWP) of 0.115. All toxicological studies on HFC-134a have been completed including one-year inhalation study with favorable results. Wilson et al. conducted a detail study and reported the thermodynamic properties of HFC-134a. The thermodynamic properties of HFC-134a are very much similar to CFC-12. Normal boiling point (NBP) of HFC-134 a is – 26.8 deg. C which is very near to normal boiling point of CFC-12 (−29 .8 deg. C). McLinden considered the use of HFC-134a to be the most realistic refrigerant to CFC-12. Thermal conductivity and viscosity were measured over a temperature range temperatures and reported by Shank land et al. Lot of research works were carried out on different aspects of using HFC134a as a real substitute to CFC12. The use of oil in HFC 134 a systems requires a very stringent quality control. Some researchers reported that HFC-134a has been proved as a fully reliable refrigerant in retrofitting CFC-12 systems including centrifugal chillers, semi-hermetic reciprocating and screw compressors and HFC-134a is not the most suitable option for the hermetic systems.

Boot in his paper "Overview of Alternatives to CFCs for Domestic Refrigerators and Freezers" concluded that alternatives beyond HFC-134a must be considered owing to the inefficiency of HFC-134a when used in a refrigerator. Vineyard et al. performed tests with HFC-134a in a standard household refrigerator and concluded that HFC-134a consumes more energy than that of CFC-12. From the thermodynamic data, it can be estimated that HFC-134a has a lower capacity and operates at lower suction and higher discharge pressure than CFC-12 for the same evaporating and condensing temperatures. Based on this information, a larger compressor would be necessary to achieve capacities equivalent to those obtained with CFC-12.

9. Hydrocarbons as refrigerants

Hydrocarbon refrigerants, which do not contain any halogen atom, are promising substitute for CFC-12. Hydrocarbons are environmentally safe, efficient, and technologically reliable refrigerants and insulation foam-blowing agents. Hydrocarbons are naturally occurring substances formed from fossilized plant matter, and found throughout the world as oil and natural gas. Lower paraffin's such as propane, butane, and isobutene were successfully used as refrigerant before the advent of chlorofluorocarbons. The thermodynamic properties of the hydrocarbons are much better than any of the other alternatives known. (Refer the **Table 4**). Hydrocarbons are available at low price all over the world and are compatible with commonly used lubricants and materials of construction used in refrigeration systems. Hydrocarbons are relatively cheap to produce and they are readily available in most parts of the world.

The need to find substitutes for CFCs during the 1990s has led refrigeration industry back to using hydrocarbons which have no impact on the ozone layer and insignificant contribution to global warming. Since 1992, hydrocarbon refrigeration has become the technology of choice in many domestic markets in Western Europe. In Germany, 100 per cent of the industry has already converted to hydrocarbon technology. All of the major European companies, such as Bosch/Siemens, Electrolux, Liebherr, Miele, Quelle, Vest frost, Whirlpool, Bauknecht, Foron, and AEG are selling hydrocarbon refrigerators. They are available in many sizes, and a wide variety of models, including some with no-frost freezer compartments. There are over 100 different hydrocarbon refrigerator models on the European market.

The main drawback of these refrigerants is their high inflammability, which has prohibited their use. But modern innovations have greatly improved the safety of hydrocarbon technologies. Besides due to very low charge of hydrocarbons (the amount of propane or butane in a domestic hydrocarbon refrigerator is only 40 to 60 g equivalent to two to six cigarette lighters, depending on the size of the refrigerator) in small capacity refrigeration system inflammability does not present a problem The present level of technological development and safety measures available have made it possible to use hydrocarbons as working fluids in domestic refrigerators. Most consumers in Southern countries are already familiar with hydrocarbons in the form of LPG (liquid petroleum gas a propane and butane mix), as it is widely used for cooking in the home. According to Pearson of Star Refrigeration over 50 million refrigerators using hydrocarbon as a refrigerator have been produced and not a single accident due to flammability was reported. The main motivation to adopt hydrocarbons in spite of their high inflammability is their being **eco-friendly**. The hydrocarbons do not destroy ozone layer and their GWPs are hundred times lower than that of CFCs and ten times lower than other CFC substitutes. Among hydrocarbon pure fluids, propane and isobutene are finding much attention as a substitute to CFC-12 in recent years [14].

9.1 Propane (HC-290)

Propane has been tested in the small capacity refrigeration systems. Propane can be considered as an alternative for HCFC-22.The high latent heat requires low refrigerant charge in the system. The performance of propane is comparable to CFC-12 and HCFC-22 and considered to be better than HFC-134a. Granryd et al. designed a heat pump prototype with propane as refrigerant as an alternative to HCFC-22. Use of HFC 134 (a) will be decomposed as acids and poisonous substances, which is worse than the use of CFC. So use of natural refrigerants like hydro carbons was suggested by Lorentzen [6].

9.2 Cyclopropane (HC-270)

Cyclopropane is also a promising refrigerant for domestic refrigerators. Kim et al. carried out simulation and experimental studies using HC-270 as a refrigerant in a single evaporator refrigerator and found encouraging results with respect to energy consumption. It was reported that cyclopropane results in lower energy consumption by 6 to 7% and 17% increase in volumetric capacity as compared to CFC-12 single evaporator refrigerator.

9.3 Isobutane

Isobutane exhibits a higher normal boiling point (−11.85 deg. C) compared to CFC-12 (−29.8 deg. C) and requires about 80% larger displacement volume in the compressor than CFC-12 to obtain an equivalent cooling capacity. Ray Riffe et al. investigated the use of isobutane as a refrigerant in refrigerator/ freezer with the use of dual cycle (D.R, 1995). His conclusions were

- For same refrigeration capacity, the amount of charge required by isobutene is 50% less than that of CFC-12.

- In order to achieve the same cooling capacity as that of CFC-12, compressor is to be replaced with a larger displacement compressor.

- Suction pressure (gauge) is slightly negative.

- Noise level of Isobutane refrigerator is much lower in comparison to CFC-12, and Isobutane results are quite encouraging.

From the above discussion it is quite apparent that there are **limited numbers of pure fluids that can function as substitute for CFCs.** mixing of refrigerants allow adjustments or tuning of the most desirable properties to provide suitable alternatives. Of course, other properties are also altered and thus the craft of developing a mixture is to obtain a final fluid with all desirable properties in the operating range. Mixtures provide a flexibility of modulating the capacity by varying the composition of the constituents. Refrigerant mixtures are solutions, i.e. they have constituents, which are equally dispersed and cannot be mechanically separated. There are three categories of mixtures.

9.4 Azeotropes

Azeotropes are also known as constant boiling point mixtures. This class of refrigerant mixture **behaves as if it were a single component during its phase change** (In the phase change the proportion of each constituent in the new phase is the same as in the original phase). Azeotropic mixtures usually behave like a fluid. This property is of great use when the refrigerant mixture leaks and it has to be recharged. However, the recharging of Non Azeotropes is much more complex [15]. The advantage of using Azeotropes as refrigerants is that, during leak in one of the components, it does not change the composition of the remaining refrigerant. Since the composition of an azetropic mixture is produced is a role of temperature, no true azeotropic mixtures be present in refrigeration. But the composition change of the azeotropic mixtures is very small. Azeotropic mixtures have been very commonly used in refrigeration, such as cold storage. Refrigerant R 502 is a mixture of R 22 and R 152 (a). R 410 (a), is a mixture of R 32 and R 125 is an important non-ODS azeotropic mixture used in place of R 22.

9.5 Near azeotropes

Azeotropes rigorously exist at only one composition for given temperature and pressure. However for all practical purposes, minor deviations are acceptable for many refrigerant systems. For this broader range the term 'Near Azeotropic Mixture' (**NEARM**) is used. The advantage of utilizing this category is that many more possible alternatives become available.

9.6 Non-azeotropes

Non – Azeotropes are also called as zeotropes. A zeotropic blend is also a combination of two or more components. The components of this refrigerant have diverse boiling points. These components will evaporate and condense at unlike temperatures. In order to fully understand zeotropic blends we must also understand fractionation and glide [16].

Non-Azeotropes, which change their composition continuously during phase, change. These mixtures do not possess a sharp boiling point but boil over a range of temperature. This feature is intimately tied to the improvement of system's efficiency if appropriate hardware changes are made to the system heat exchangers. The change of temperature with phase change is called Temperature Glide.

When heat transfer fluids exchange heat with Non Azeotropic Refrigerant Mixtures (NARMs) in a constant current flow mode, the thermodynamic irreversibility can be reduced by matching the temperature glide (NARM side) against the temperature drop (Heat Transfer Fluid side) resulting in an increase in the coefficient of performance.

As already stated, because of limitation of single fluids as alternative to CFC-12 and also the flexibility in modulating the capacity by varying the composition of the constituents, mixtures are emerging as a viable solution to CFC-12 alternatives.

10. Mixtures of propane and isobutane

Propane and Isobutane will combine pretty happily as both are non-polar. Butane molecules will be broken and are replaced by the molecules of butane and Propane molecules. Propane-isobutene mixture has the benefit of modulating the capacity to permit their use with compressor designed for use with CFC-12. The binary mixture ratio can conceivably be designed using boiling point as a guideline such that the disparate requirements of the freezer and the fresh food compartment in domestic refrigerators can be balanced. The normal boiling point temperature for HC-290/ HC-600a mixture (each 50% by mass) ranges from −32 deg. C to −24 deg. C which is very close to normal boiling point of CFC-12 [17].

The mixture in comparison to CFC-12 possesses very high latent heat of vaporization and low value of density (1/3 of CFC-12), which makes the mixture attractive because of its low charge requirement and circulation rates. The charge levels are approximately 40% that of CFC-12. One of the important advantages of

R-600a/ R-290 blends is that it is compatible with mineral oils and commonly used materials for manufacturing of refrigeration systems and requires minimal changes in the refrigeration systems. The mixtures do not contain any halogen atom and hence the possibility of forming acids in the presence of moisture is eliminated [6].

This is also a factor for better stability of these mixtures. **Table 4** shows the thermodynamic properties of few hydrocarbons and CFC-12. It follows from **Table 4** that pure hydrocarbons cannot be exact alternative to CFC-12. Meyer conducted experimental investigation with the use of hydrocarbon refrigerants for domestic refrigerators. It was found that an unchanged CFC-12 single temperature refrigerator had lower energy consumption with a refrigerant mixture of 50% propane and 50% isobutene compared to CFC-12. Based on these encouraging results FORON decided to develop single temperature refrigerators using hydrocarbon mixtures as refrigerants. This was the beginning of use of hydrocarbon mixtures as refrigerants in domestic refrigerators. Domanski et al. carried out a study of hydrocarbon refrigerants for residential heat pump systems and concluded that these mixtures were promising substitute. He examined the various benefits and discussed the Rankine Cycle System and component design issues and limitations when using a wide range of single component refrigerants, hydrocarbons, with the help of the READER code for residential size heating and cooling systems (Domanki, 1994).

Liu et al. (1994) conducted an experimental test with propane-isobutane as a drop in substitute in a domestic refrigerator/freezer unit. They kept all hardware components of the refrigerator/freezer as that of CFC-12 except the capillary tube, which was, lengthen to control the flow rate. They concluded that highest savings of 6.5% were achieved with a blend of 70% HC-290 and 30% HC-600 with a charge of 70 g.Kruse performed a theoretical evaluation of hydrocarbons and its mixtures as refrigerants in refrigerators, unitary air conditioners and heat pumps. He concluded that in general, **hydrocarbons have an inherent possibility of lower energy consumptions.** He also mentioned that refrigerators with hydrocarbons mixtures as refrigerant shows further possibility of energy improvement if they are designed by using Lorenz-Meutzner cycle. The major drawback of these mixtures, like other hydrocarbons, is the high inflammability. But the charge is only 1/3 to ½ that of CFC 12. Owing to the fact, inflammability does not pose a problem. However, proper safety precautions should be taken at different stages of manufacturing, charging and operating (Cichong Liu, 2016).

When the Montreal Protocol has phased out CFCs in the developed countries and has a regulatory regime for the phase out of HCFCs, the problem still is a major problem in developing countries. Awareness is to be created among the common people to save the planet earth. Developing countries should phase out ozone depleting chemicals in a very careful manner.

Global market will continue to develop & introduce new refrigerant chemicals. Statutory requirement is necessary to control the UN –organized sector who are dealing in HVAC industry.

11. Economic impact of the alternative refrigerants against cfc refrigerants

Every human being or a machine has a value. Mechanical machines are designed, fabricated and manufactured with different materials, and so cost is incurred for the manufacture of all the machineries. Almost all the residential air conditioners are vapor compression systems, which uses compressors. Compressor is the main component in any air conditioning system, and so it is the costliest component in an air conditioning system. Other components and the type of refrigerants used in the air conditioning systems will also contribute significantly to the total cost of the air conditioning system. The cost of one ton window air conditioner as an example is shown in the below **Figure 2**.

11.1 Cost analysis

Cost incurred in the design and fabrication of constructal designed window air conditioner is shown in the below **Figure 3**. Here, the compressor was replaced with a pump and so the major part of the cost is reduced. Also the refrigerant materials, phase change material and water were industrial waste from the refineries and natural resource respectively.

CFC type refrigerants were used in window air conditioners, and now it is is completely phased out and replaced with HFC type HC (Hydro carbon) type refrigerants. It is difficult to fill in these refrigerants in the same system which used CFC type refrigerants. The system needs lot of design changes and the air conditioning service technicians carelessly releases the harmful CFC gases into the

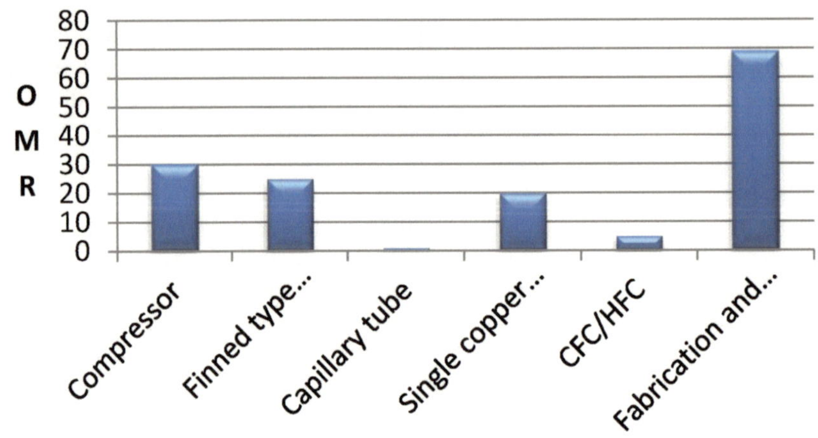

Figure 2.
Cost of one ton of window air conditioner.

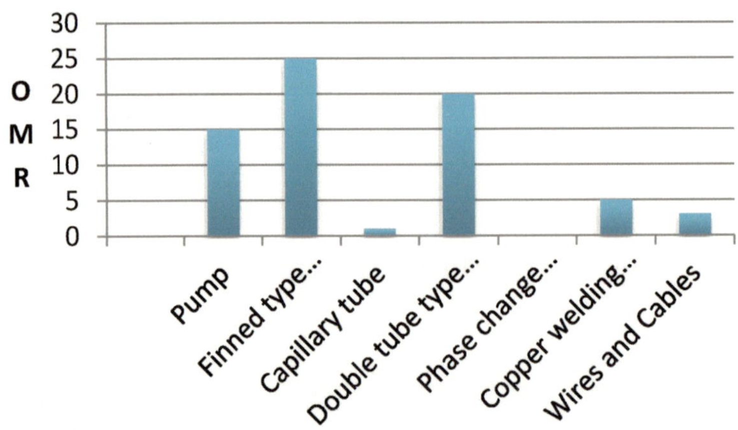

Figure 3.
Cost of Constructal designed window air conditioner.

Sl. No	Refrigerant name	Cost in INR	Cost in OMR	Cost USD
1	R134 (a)	Rs 390/Kg	R.O 2.02/Kg	USD 5.96/Kg
2	R 22	Rs 380/Kg	R.O 1.97/Kg	USD 5.13/Kg
3	R 404	Rs 400/Kg	R.O 2.08/Kg	USD 5.39/Kg
4	R 410 (a)	Rs 350/Kg	R.O 1.82/Kg	USD 4.72/Kg
5	R 32	Rs 395/Kg	R.O 2.05/Kg	USD 5.33/Kg
6	R 600 (a)	Rs 1000/Kg	R.O 5.19/Kg	USD 13.49/Kg
7	R414 (b) Hydrocarbon Blend	Rs 1250/Kg	R.O 6.49/Kg	USD 16.86/Kg
8	Phase Change Material	Rs 30/Lt	R.O 0.16/Kg	USD 0.40/Kg

Table 6.
Cost of refrigerant materials.

atmosphere. So, it is a good opportunity to retrofit the old window air conditioners with non CFC refrigerants. The total cost incurred for the fabrication of constructal designed window air conditioner was found to be around R. O 70/−, which was very much lesser than the normal vapor compression type window air conditioner. Cost can be further reduced by using the thrown away air conditioners and using the condenser and evaporator coils.

11.2 Cost of alternate refrigerants

Cost of refrigerants as per the current selling prices in the refrigeration and air conditioning markets in Oman and India are presented below. Phase change materials can be extracted from the industrial waste from the petroleum refineries. Hence, there is a huge potential to recycle the industrial waste and thereby contributing indirectly to safe environment (**Table 6**).

12. Summary

Refrigerants are the key substances used in all the conventional refrigeration and air conditioning systems. Refrigerants rub though out the system and removes heat by changing its phases during the course of operation. Use of refrigerants evolved from the day of first refrigeration and air conditioning equipment. Air, carbon dioxide, ammonia, sulfur dioxide were used as refrigerants during the early day air conditioning systems. But because of the need in very low temperature applications and human comfort conditions, different artificial refrigerants were came into use. Significant artificial refrigerants are chloro flouro carbons (CFC), Hydro flouro carbons (HFC). Though the temperature produced by these refrigerants are very good for wide applications in residences and industries. But they contribute indirectly to the ozone layer damage and global warming.

To ensure the safety of environment, use of CFCs are banned and to be phased out completely as per the guidelines by Montreal Protocol. Even though, phasing out of CFCs and HFCs are not happening in developing countries, and so the threat to the environment continues. This chapter, introduces the use of phase change materials) PCM) as a potential refrigerant in air conditioning systems. Since phase change materials are thrown as waste byproducts from the petroleum refineries, use of PCMs in air conditioning systems will be additional contribution for the safety of environment.

Nomenclature

cfc = chloro flouro carbon.
hc = hydro carbon
ac = air conditioner
k_{eff} = effective thermal conductivity
q_{pc} = heat flux
r_c = radius of centroid
α = void fraction
μ = viscosity of fluid
ρ = density of fluid
mmtcde = million metric tons of carbon dioxide equals
gwp = global warming potential

Author details

Selva Pandian Ebenezer
College of Engineering, National University of Science and Technology, Muscat, Sultanate of Oman

*Address all correspondence to: selvapandiane@gmail.com

References

[1] M. F. Lunt *et al.*, "Reconciling reported and unreported HFC emissions with atmospheric observations," pp. 1–5, 2015, doi: 10.1073/pnas.1420247112.

[2] M. R. Qader, "Electricity Consumption and GHG Emissions in GCC Countries," pp. 1201–1213, 2009, doi: 10.3390/en20401201.

[3] R. Saidur, S. Ma, H. H. Masjuki, and M. Y. Jamaluddin, "Greenhouse Gas Emissions From Refrigeration," no. x, pp. 533–552.

[4] T. M. Protocol *et al.*, "History of Chlorofluorocarbons Influence on Depletion of the Ozone Layer and Global Warming The Mechanism of the Ozone Layer Depleted by Ozone Depleting Substances The Mechanism of Global Warming Caused by Three CFC Alternatives," 1987.

[5] U. Langematz, "Stratospheric ozone : down and up through the anthropocene," *ChemTexts*, vol. 5, no. 2, pp. 1–12, 2019, doi: 10.1007/s40828-019-0082-7.

[6] J. H. Koh, Z. Zakaria, and D. Veerasamy, "Hydrocarbons as Refrigerants — A Review," vol. 34, no. 1, pp. 35–50, 2020.

[7] M. Kuta, D. Matuszewska, and T. M. Wójcik, "Reasonableness of phase change materials use for air conditioning – a short review," vol. 33, pp. 0–7, 2017.

[8] N. Chaiyat and T. Kiatsiriroat, "Energy reduction of building air-conditioner with phase change storage," *Int. J. Appl. Eng. Res.*, 2014, doi: 10.1016/j.csite.2014.09.006.

[9] J. Virgone, Experimental assessment of a phase change material for HAL *Id : hal-00471155*, no. October. 2009.

[10] M. Imran Hossen Khan and H. M. Afroz, "Effect of Phase Change Material on Performance of a Household Refrigerator."

[11] M. A. Boda, R. V Phand, and A. C. Kotali, "Various Applications of Phase Change Materials: Thermal Energy Storing Materials," *Int. J. Emerg. Res. Manag. &Technology*, vol. 6, no. 4, pp. 2278–9359, 2017.

[12] J. Yen Chou, "Phase Change Materials for Energy Efficient for Energy Efficient Housing Applications," 2008.

[13] European Commission, "Directive 2006/40/EC of the European Parliament and of the Council of 17 May 2006 relating to emissions from air-conditioning systems in motor vehicles and amending Council Directive 70/156/EEC," *Off. J. Eur. Union*, no. 161, pp. 12–18, 2006.

[14] M. Sruthi Emani and B. Kumar Mandal, "The Use of Natural Refrigerants in Refrigeration and Air Conditioning Systems: A Review," *IOP Conf. Ser. Mater. Sci. Eng.*, vol. 377, no. 1, 2018, doi: 10.1088/1757-899X/377/1/012064.

[15] S. Benhadid-Dib and A. Benzaoui, "Refrigerants and their environmental impact substitution of hydro chlorofluorocarbon HCFC and HFC hydro fluorocarbon. Search for an adequate refrigerant," *Energy Procedia*, vol. 18, no. December, pp. 807–816, 2012, doi: 10.1016/j.egypro.2012.05.096.

[16] S. S. Jadhav and K. V Mali, "Evaluation of a Refrigerant R410A as Substitute for R22 in Window Air-conditioner," *IOSR J. Mech. Civ. Eng.*, pp. 2278–1684, [Online]. Available: www.iosrjournals.org.

[17] S. O. Banjo *et al.*, "Experimental analysis of the performance characteristic of an eco-friendly HC600a as a retrofitting refrigerant in a thermal system," *J. Phys. Conf. Ser.*, vol. 1378, no. 4, 2019, doi: 10.1088/1742-6596/1378/4/042033.

Chapter 6

Impact of Working Fluids and Performance of Isobutane in the Refrigeration System

Solomon O. Banjo, Bukola O. Bolaji, Oluseyi O. Ajayi and Olatunde A. Oyelaran

Abstract

The effect of heat transfer medium (HTM) on the environment is associated with ozone layer depletion and global warming. The role of HTM (working fluid) in the heating and air conditioning industries is paramount, which cannot be underestimated. The conventional refrigerant has been predominantly used over decades due to their thermodynamic properties. However, hydrocarbon refrigerants such as isobutane are considered substitutes because they have negligible global warming potential and zero ozone depletion. That makes it eco-friendly among other existing refrigerants. The investigation of the refrigeration system's performance characteristics required consideration for the coefficient of performance, refrigerating effect, and the compressor work; this enables the determination of the system's efficiency without any assumption. Another factor that suggests a better absorption of refrigerant (working fluid) into a refrigeration system is an increase in the coefficient of performance (COP). The effect will cause a reduction in the rate of energy consumption by the compressor. The result shows that the system's coefficient of performance when using R600a was 27.1% higher than when working with R134a, with an energy reduction of 23.3%.

Keywords: coefficient of performance, refrigerating effect, global warming potential, working fluid, ozone depletion potential

1. Introduction

At the beginning of the eighteenth century, natural ice was exploited for domestic and commercial purposes, such as food preservation. Also, in the 1800s, there was a discovery of volatile liquids that could condense by applying compression and cooling [1]. The combination of these two inventions results in the fabrication of the household refrigeration system, which has worldwide applications. Since the nineteen centuries that vapor compression system (VCS) has been invented, its practical implementation has cut across various fields, including preservation of food and vaccine, heat ventilation and air-conditioning for human comfort, and storage of farm produce, and industrial processing. Preservation became essential to expand the product's shelf life, which enables the quality in terms of physical properties that include color, texture, and flavor. More so, the refrigeration process, among other food storage methods, has been proved to be most effective, dependable, desirable, and applicable worldwide [2–5]. Over two decades ago, thermal systems have found increased application in the tropical region, and they consumed a massive amount of electrical energy. Furthermore, contrary to the excessive energy consumption rate exhibited by the refrigeration system (RS), the refrigerant plays a vital role in the overall performance of a vapor compression system [6].

The heat gained by the refrigerant in VCS flows in the direction of decreasing temperature, from the high-temperature region to the low-temperature region. Also, there are assumptions required for an ideal vapor compression system to occur. These include:

- the heat rejects to the immediate surroundings are ignored.
- there are no frictional pressure drops.
- working fluid flows at constant pressure in the condenser and evaporator, which serves as the refrigeration system's heat exchangers.
- process of compression is isentropic the irreversibility process, within the essential components such as compressor, condenser, and evaporator, are ignored [7, 8].

More so, there are four processes involved in establishing low temperature in a refrigeration system was explained and represented by a pressure-enthalpy (p-h) diagram, as shown in **Figure 1**.

1.1 Isobaric heat absorption process

From **Figure 1**, point 4 to 1, the working fluid has low pressure and temperature in a vapourized form at constant pressure and temperature in the evaporator. The latent heat absorbed in the evaporator by the working fluid is expressed in Eq. (1). The phase change occurs as the refrigerant isentropically compressed by the compressor and moves to the next stage.

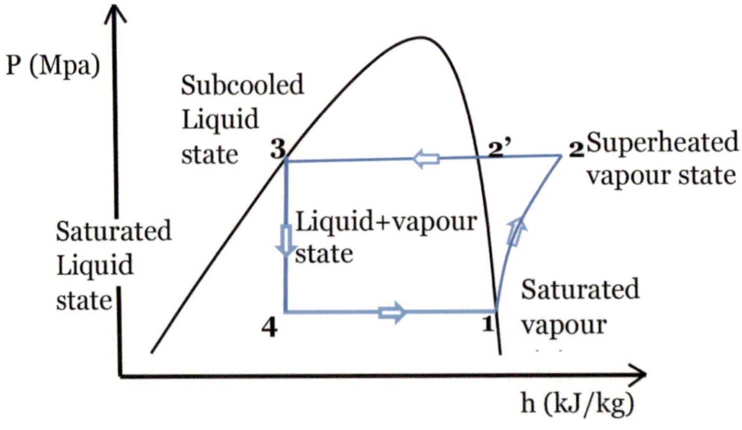

Figure 1.
P-h diagram of a vapor compression cycle.

1.2 Isentropic compression process

The isentropic compression takes place in the compression line. The vaporized refrigerant from the evaporator is compressed isentropically from stages 1 to 2 by the compressor. Work done on working fluid by the compressor is given in Eq. (2). The compression force increased the refrigerant's pressure and temperature, resulting in the rise of pressure and temperature, as shown in **Figure 1**. And refrigerant attains its superheated state at point 2.

1.3 Isobaric heat rejection process

The superheated refrigerant was de-superheated at constant pressure through the compressor outlet temperature to the condenser temperature at point 2′. Then, condensation occurs due to the natural air that inter-phases with the condenser's extended surface, resulting in heat rejection from the refrigerant. Thus, reducing the temperature by condensation at the condenser enables the refrigerant to attain a sub-cool state at point 3 at constant pressure and constant temperature along the 2′ and 3 as shown in **Figure 1**, and the heat loss is expressed in Eq. (5).

1.4 Isenthalpic expansion process

In this isenthalpic process, the pressure at the upstream consistently higher than the pressure at the low stream. The capillary tube (CT) is used in practice to replace the expansion or throttling valve in a vapor compression refrigeration system (VCRS). This component aims to drastically reduce the pressure of the refrigerant that throttles through stages 3 and 4. It is assumed that there is no heat gain or loss because the process is adiabatic, that is, h3 = h4, as displayed in **Figure 1**. Therefore, the refrigerant throttled down the capillary tube and moved to the evaporator inlet at point 4.

In cooling systems, emphasis is always on domestic refrigerators and air-conditioning (AC) systems. There are various types of refrigeration and AC systems. RS can be classified base on the kind of energy input and the refrigeration process as:

2. Natural refrigeration

2.1 Art of ice making by nocturnal cooling

The art of making ice by nocturnal cooling is a common method of producing ice in India. It involves keeping a thin layer of water in a deep earthen vessel such as a tray with compacted hay of about 0.3 m thickness, which serves as an insulator, and as the tray was exposed to cool air, water emits heat by radiation to the stratosphere, almost at −55°C; the water in the trays turns to ice [9].

2.2 Evaporative cooling

The evaporative cooling method has been adopted in India for many centuries to obtain cold water in summer by storing the water in earthen pots. The water penetrates through the pores of the earthen vessel to its outer surface where it evaporates to the immediate surrounding, and absorb its latent heat from the vessel, which cools the water. In recent days, desert coolers are used in hot and dry regions to provide cooling in summer [10, 11].

3. Non-natural refrigeration

3.1 Vapor compression refrigeration systems

See Section 2.

3.2 Thermoelectric refrigeration systems

The thermoelectric refrigeration system (TRS) uses electricity to provide cooling effect utilizing the principle of two dissimilar metals to generate emf. There are two basic methods in obtaining this operation, which include Peltier and Seebeck effect.

3.2.1 Peltier effect

This requires the flow of electric current through two dissimilar conductors, and the junction of the metals (conductors) either emit or absorb heat, which depends on the current that flows across the junctions. However, the flow of electric current is proportional to the heat gain or loss at the junction [12].

3.2.2 Seebeck effect

In this process, direct heat is converted to electricity at the junction of various conductors (wire). This requires the generation of voltage along the wire that is subject to the temperature gradient and it results in thermoelectric power (see **Figure 2**) [13].

3.3 Vapor absorption refrigeration systems

The vapor absorption refrigeration system (VARS) comprises the same processes in the vapor compression system (VCS) such as compression, expansion, condensation, and evaporation. In the vapor absorption system, the working fluid used is ammonia, water, or lithium bromide. The refrigerant produces a cooling effect in the evaporator, and heat dissipates to the atmosphere through the condenser. The main difference between the two systems is the suction and compression of the refrigerant in the refrigeration cycle. In the VCS, the compressor sucks the refrigerant from the evaporator and compresses it to high pressure. The compressor also enables the flow of the refrigerant through the whole refrigeration system. In the VAS, the process of suction and compression is carried out by two

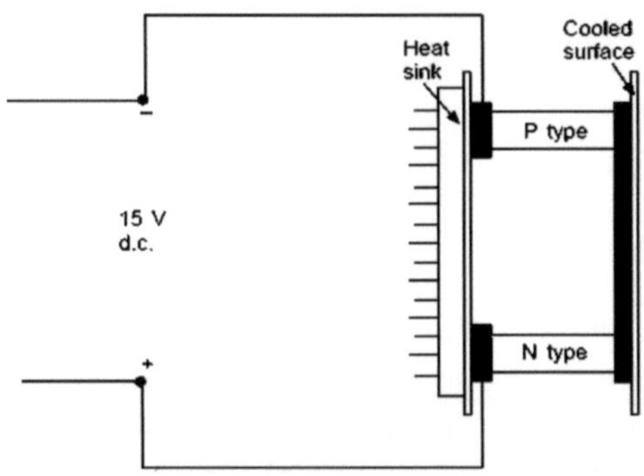

Figure 2.
Thermoelectric refrigeration systems.

DOI: http://dx.doi.org/10.5772/intechopen.99121

different devices called the absorber and the generator. The absorber and generator replace the compressor in the VAS (**Figure 3**) [14].

3.4 Vortex tube systems

This cooling method is obtainable by the compressed air entering the vortex tube at high pressure through a tangential nozzle, which serves as an accelerator for the flow of air and creates high rotational speed and velocity. The air possesses a whirling motion that resulted in cold and hot air, which discharge through the cold and the hot pipes attached to the vortex tube [15].

3.5 Steam ejector refrigeration system

In a steam ejector refrigeration system (SERS), high-pressure steam of 10 bar is employed at a velocity of about 1200 m/s. The water evaporates at 4–7°C for air-conditioning duty was obtained with a steam ejector at low pressures ranging between 8 and 22 mbar. However, SERS' coefficient of performance is lesser than that of the vapor absorption system, and this placed a limit in its application where a large quantity of steam is required (**Figure 4**) [16].

3.6 Air expansion refrigeration systems

There is a temperature rise when air is compressed. Thus, the temperature drops when the air is allowed to do work while expanding and is obtained by sensible heat only, which is the basis of the air liquefaction process. The main application for the expansion refrigeration system cycle is in the pressurization of aircraft and air

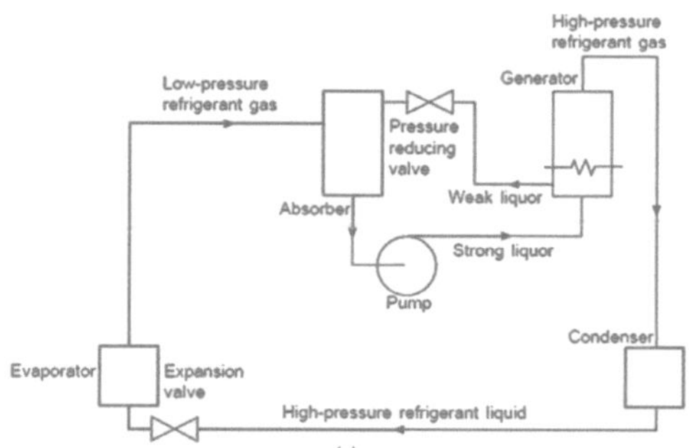

Figure 3.
Vapor absorption refrigeration systems.

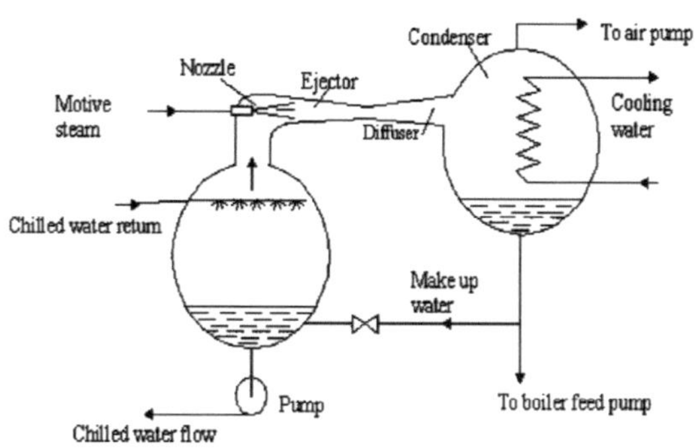

Figure 4.
Steam jet refrigeration system.

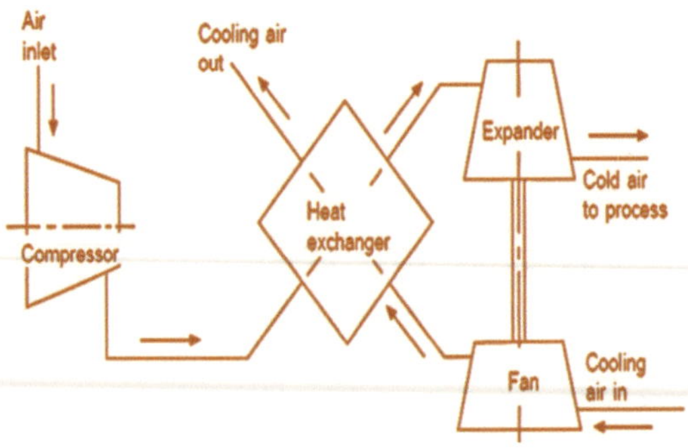

Figure 5.
Air or gas expansion refrigeration systems.

conditioning. Also, the turbines used compression and expansion turns at very high speeds to achieve the actual pressure ratios, and as a result, it is noisy. Likewise, the coefficient of performance is lower compare to other refrigeration systems. The normal cycle uses the expansion of the air to drive the first stage of compression, so retrieve some of the input energy (**Figure** 5) [17].

4. Type of air conditioning system

4.1 Domestic air conditioning system

4.1.1 The window-type air conditioner

This type of air conditioner has all the components assembled in one compartment. It is often inserted through the wall of a building, but due to the configuration and installation, the hole carved on the building wall weakens the structure over a while. It also provides an opportunity for external factors such as rats and snakes to enter through the small openings on the wall. Though, it is readily available and cheap [18, 19].

4.1.2 Split type air conditioner

The split type of air-conditioning is an upgrade of the window air conditioner with two separate. The indoor unit mainly contains the evaporator, vent, evaporator motor, filter, electric panel, and the outdoor comprised of the compressor, condenser, capacitor, fan blade, electrical panel, accumulator, and expansion valve. This air conditioner is more effective in terms of performance and does not require a sizeable expanse of the hole in the building. Though it is more expensive, the maintenance cost is higher than the window air conditioner [20].

4.2 Industrial air conditioning

4.2.1 Standalone air conditioner

It is an industrial-scale air conditioner that comes with different tons in terms of duty. Also, it has two units, indoor and outdoor. This type of air conditioner is higher in capacity compared to the window and split air conditioners. It comes in various sizes depending on the capacity and manufacturer's design. Its limitation is that it occupies space and expensive [19, 21].

4.2.2 Central air conditioner

The central air-conditioner is an improvement of the standalone air conditioner. This air-conditioner found its operations in an extensive environment such as banks, shopping malls, Churches, and Conference centers. It is most effective and works with a network of the ducting system to enable even distribution of cool air across the desired area to be cooled even though it is costly in terms of running cost and maintenance [22, 23].

5. Low-temperature technology

Low-temperature technology is achievable through the refrigeration processes, particularly in the vapor compression refrigeration system (VCRS). This process requires heat transfer from the refrigerating space through a heat exchanger known as the condenser. The heat moves to the immediate surroundings, thereby lowering the heating space's temperature and increasing the surrounding temperature [24, 25]. However, the application of VCRS has become a worldwide phenomenon. The VCRS is used to protect perishable items and provide human comfort [26–29]. In designing a VCRS, the refrigerant (working fluid) choice is crucial as it preempts the system's cost, dependability, and safety. Considering various environmental challenges such as global warming and ozone layer depletion, the appropriate selection of refrigerant is required due to the atmospheric greenhouse effect and stratospheric ozone depletion caused by refrigerant emission [21]. The refrigerant is selected based on their safety, thermo-physical and thermodynamics properties, and economic factor [30].

In 1850, ethyl ether became the most prominent refrigerant used in a commercial refrigeration system. Other natural refrigerants such as ammonia, water, carbon dioxide, and gasoline were exploited as heat transfer mediums. Still, they were replaced with sulfur dioxide and methyl chloride because of incompatibility with the system materials. Moreover, at a low cost and good thermodynamic properties, ammonia has an increased coefficient of performance. However, it is toxic, preventing it from being used for domestic purposes [31–33]. Therefore, the search to have better-working fluid with excellent thermal properties used in a refrigeration system has prompted the discovery of halocarbon refrigerants. The halocarbons are synthetic refrigerants, which include chlorofluorocarbon and hydrochlorofluorocarbon refrigerants (HCFCs). They were better alternatives for natural refrigerants because of their high thermal efficiency [34–37]. The chlorofluorocarbon (CFC) refrigerant was discovered in the 1930s with zero explosion risk but later disadvantaged the ozone layer [30]. In 1996, the developed nations successfully phased out the CFC, while the developing countries agreed to cease its application in 2010 [38]. The HCFCs were found as the best short-term alternative replacement for chlorofluorocarbon, but with a global warming potential of 1810. The use of HCFCs will be limited to 2020 and 2030 in developed and underdeveloped countries. Likewise, Du Pont's chemical manufacturer declared that the HCFC substances' production should be ceased but would persist in making it available for existing equipment until their expiry period [39–41].

The emergence of hydrofluorocarbon (HFCs) refrigerants with appropriate thermo-physical and thermodynamic properties have provided opportunities for extensive scale usage at the consumer and commercial levels since the 1990s. This rapid growth and application of HFCs in domestic and mobile refrigeration systems require prompt attention due to their negative effect on global climate [42, 43]. However, hydrofluorocarbon refrigerant (R134a) has zero ODP with a high GWP of 1430, which suggests it is a working fluid that threatens the immediate surroundings. The international regulatory bodies (Kyoto and Montreal protocols) have called for the banning of pure fluid, posing a threat to climate and environment due to high global warming and ozone depletion [44–46]. But the reduction rate of the pure fluid is subject to review. More so, some limiting factors disannulled halocarbon refrigerants' application in the heating and air-conditioning industries, such as the enormous energy consumption rate and increased global warming [47]. Furthermore, the Kigali amendment (KA) adoption in 2016 created a platform to generate a phase-out schedule for HFC refrigerants towards the next decades, and if this is achieved, it will contribute positively to the Paris agreement (PA) and to the United Nation Framework Convention on Climate Change (UNFCCC) adopted in 2015, which focuses on keeping the global temperature to less than 2 oC with the enforcement by Nationally Determined Contribution (NDC) for greenhouse gas (GHG) [48–51]. The consistent climate change prompted research towards discovering and applications of eco-friendly refrigerants in the heating and air-conditioning systems that reduce GHG and enhance energy saving [52].

Halogen-free refrigerants (HFRs) are found naturally and have been extensively discovered as an alternative to halocarbon refrigerants in the refrigeration system. They are also referred to as eco-friendly refrigerants with an organic composition of hydrogen and carbon atoms [21]. The European nations use HFRs instead of halocarbon refrigerants because it possesses good thermodynamic properties. Hydrocarbon refrigerant is miscible with mineral oil, which provides a smooth running of the single hermetic reciprocating compressor (SHRC). The working fluid is compatible with the refrigeration system's elastomeric materials with zero ozone depletion and negligible global warming potential. Also, the HCR refrigerant has a high critical temperature that enhances the domestic refrigerator's efficiency [53–59]. The HFR is for preserving the environment, but it serves as an energy reduction substance with high energy efficiency in the refrigeration system. Although HFRs was reported flammable [60], this proved to be invisible because of various factors that must be attained before the explosion takes place, which includes:

Refrigerant type	**Composition**	**GWP**	**ODP**	**Safety group**
HCFC 22 refrigerants				
R22	CHClF2	1810	0.055	A1
R124	CHClFCF3	609	0.022	A1
R142b	CClF2CH3	2310	0.065	A2
HFC refrigerants				
R134a	CF3CH2F	1430	0	A1
R152a	CHF2CH3	124	0	A2
R125	CF3CH2F	3500	0	A1
R143a	CF3CH3	4470	0	A2L
R32	CH2F2	675	0	A2L
Hydrocarbon refrigerants				
R600a	CH(CH3)2CH3	3	0	A3
R290	CH3CH2CH3	3	0	A3
R1270	CH2CH2CH2	3	0	A3
R170	CH3CH3	6	0	A3

Table 1.
Properties of refrigerants [21, 71].

i. there must be the presence of an ignition source

ii. a surface temperature that exceeds 440°C

iii. release of hydrocarbons is essential, which mix with the appropriate proportion of air [61].

The halogen-free refrigerant is applicable in a closed system with relatively less mass charge. It attains a high level of recognition in the '90s and is used in significant engineering systems such as water dispensers, deep freezers, air-ventilation machines, domestic refrigerators, and industrial refrigeration systems [62, 63].

The refrigerants are classified based on their chemical composition and safety [64]. Refrigerant properties are shown in **Table 1**.

6. Environmental effect and economic impact of hydrocarbon refrigerants

Hydrocarbon refrigerants (HCs) such as carbon dioxide, water, ammonia, butane, propane, isobutene, and propylene are commonly used in residential and commercial buildings to attain a cooling effect. However, the HCs refrigerants have found their heating, ventilation, and air conditioning system applications due to their advantages over conventional refrigerants in terms of global warming, ozone depletion, and energy reduction in energy consumption. HC refrigerants do not have any halogen elements in their compound formulation, which enable a safe environment and prevent climate change [59]. Most developed nations have used HCs refrigerant in their refrigeration systems, such as domestic refrigerators, cars, heat pumps, and industrial air cooling systems, etcetera.

The economic implication of hydrocarbon refrigerants is obvious, the price of the compressor that uses HCs is less compared to a compressor that uses HFC, and the mass charge of HC refrigerant is lower compared to HFC refrigerant, which is ratio 1:2 respectively [65]. The reduction in mass charge of isobutane (HC600a) was due its higher value of latent heat or high volumetric capacity [25]. Also, HC refrigerant, when in use, saves more energy compare to HFC refrigerant (see **Figure 6**), and that leads to cost reduction. Moreover, the cooling capacity of a

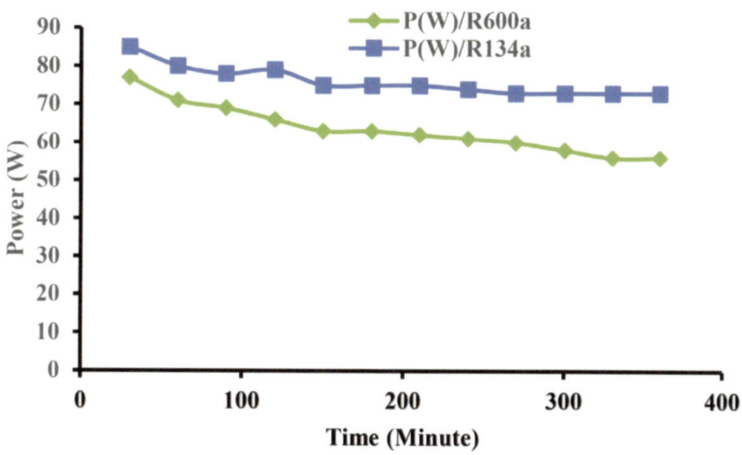

Figure 6.
Correlation between power input for R600a and R134a with time.

Compressor type		Key indicator	Economic analysis
HFC134a	HC600a		
213	186	GWP	32.6 €/t
185	158	SO_2	7.4 €/kg
41	36	NO_2	2.3 €/kg
169	148	PM10	17 €/kg
608	527	Total	

Table 2.
Economic analysis of emissions for compressor HC600a and HCF134a [66].

refrigeration system can be enhanced using the cryogenic refrigeration approach. According to [66], the economic analysis of emissions for compressor that uses HC600a and HFC134a refrigerants for 15 years lifecycle evaluation (LCE) was taken, despite of wear coefficients of the lubricant and lower friction rate the energy reduction rate was 6.54% lower when using HCF134a. The evaluation of environmental cost was based on key environmental indicators, GWP (Carbon dioxide/CO_2), SO_2, NO_2 and PM10. See **Table 2** for details.

7. Cryogenic refrigeration cooling system

Cryogenic refrigeration generates low temperatures far below the average temperature produced by a simple vapor compression refrigeration system. It is used to determine temperature range from −150°C to −273°C, and the gases associated with cryogenic are Nitrogen (N2), Oxygen (O2), Helium (He), and Hydrogen (H2), which boiled below −150°C, while other refrigerants have their boiling point above −150°C. In Engineering, the role of cryogenic includes electronics, rocket propulsion system, food preservation, nuclear engineering application, mechanical design, and biological application. Furthermore, the cryogenic approach of producing low temperature is used for commercial application and this achievable by different methods such as cascade refrigeration and multistage compressor [67, 68].

7.1 Cascade refrigeration

This system was first introduced in 1877 by Pictet, where he used for the liquefaction of oxygen (O2). There are two or three vapor compression systems involved, and for it to generate low temperature, various kinds of refrigerants were employed at the different circuits. It has application in medical and industrial systems, and one typical example is the preservation of blood in the blood bank because blood requires a temperature as low as −80°C [69].

7.2 Multistage

This type of system has higher volumetric efficiency, which results in increased compressor capacity compared to single-state compression. The cooling effect in a multistage system (MS) is obtainable by using two or more refrigerating spaces at various temperatures. The MS is represented in **Figure 7**. There is a limitation on the lowest temperature that can be attained using this method due to the usual

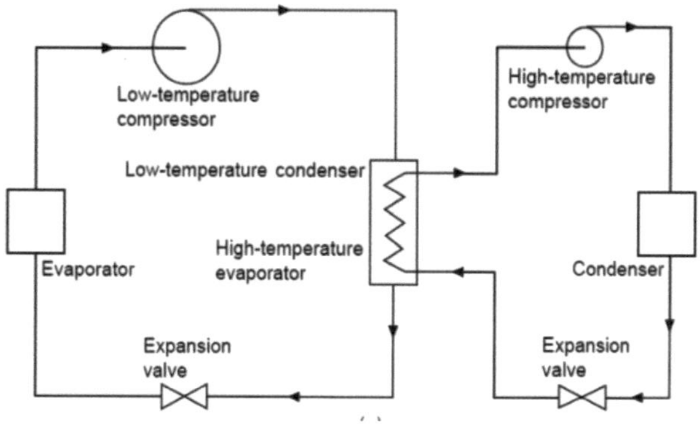

Figure 7.
Systematic diagram of the multistate system.

boiling point of the refrigerant. Thus, this system is not appropriate for the production of cryogenic temperature [70].

8. Comparison between cryogenic and refrigeration system using isobutane as refrigerant

From the description of the cryogenic refrigeration system (CRS), it visible that,

i. It can be used for commercial purposes, while a refrigeration system with isobutane is often used for domestic applications and small-scale businesses.

ii. It has a temperature range between −150°C to −273°C, while the system using isobutane does not have such a low temperature because it has a boiling point above −150°C.

iii. It can use different refrigerant fractions simultaneously, while the isobutane refrigeration system made use of one type of refrigerant at a time.

iv. In terms of size, CRS is larger than the system that uses isobutane due to its volumetric capacity and work done.

However, this study focused on the importance, properties, challenges, limitations, and refrigerants' applications in the domestic refrigeration system. The refrigeration system's heat transfer fluid (refrigerant) determines the overall performance of the system. More so, the refrigerant serves as the heat transfer medium, while the refrigerant type enables the prediction of a suitable refrigerant to be used in the vapor compression refrigeration system. This work's case study is an ideal system using isobutane (R600a) and hydrofluorocarbon (R134a) as refrigerants. And comparisons were carried out to investigate their performance characteristics.

8.1 Solubility/miscibility of lubricant

The heat transfer within the heat exchanger (H.E) of the refrigeration system is enhanced by the lubricant type used in the process. However, the solubility of the working fluid in the lubricant improves the heat transfer due to a reduction in the oil's viscosity, which increases the oil fluidity and reduces the negative impact of the heat transfer within the heat exchanger [21]. Therefore, the newly manufactured lubricant is recommended to operate a refrigeration system that uses hydrofluoro-olefins, hydrocarbon, and hydrofluorocarbon blend as working fluid is polyol-ester (POE) oil. It possessed better lubricating properties than conventional oils. Thus, the refrigerant effect becomes more visible on the heat transfer rate within the vapor compression refrigeration system as the appropriate lubricant is charged into the compressor [71].

8.2 Procedures for vapor compression refrigeration system analysis

The system has some essential components through which the thermodynamic properties were measured using various mechanical devices on the vapor compression refrigeration system. These components include a single hermetic compressor, evaporator, standard parallel tube condenser, and capillary tube, as shown in **Figure 8**.

- The vapor compression refrigeration system is positioned for an operation as displayed in **Figure 8**.
- Digital weighing balance was used to measure the refrigerant mass charge into the system.
- The temperature sensors were connected to each of the cardinal points of the refrigeration system.
- Pressure gauge was fixed to the compressor suction and discharge points.
- Power meter was connected to determine the power input to the compressor

Figure 8.
Prototype of the vapor compression system.

- A vacuum pump was introduced intermittently to trap the gas and moisture content within the system to prevent clogging

- Digital halogen leakage detector was used to detect leakages along the pipeline.

- The system was allowed to run, and data were captured at an interval of 30 minutes.

- Acquired data were computed for analysis, the performance characteristics of the VCRS that is refrigerating effect, the coefficient of performance, and the compressor work under the ambient temperature of 29°C were determined.

8.3 The basic equation for standard vapor compression system

The following expression explains the relationship between the heat input and output of a refrigeration system. The availability of a pure substance can be defined by Eqs. (1)–(6) [25, 40, 47], assuming there is an insignificant change in kinetic and potential energy across the four essential components [30].

Heat absorbed in the evaporator

$$Q_e = \dot{m}(he_1 - he_4) \text{ in kW} \tag{1}$$

Where,
Q_e = heat of evaporator.
he_1 = specific enthalpy of vapor existing evaporator in kJ/kg.
he_4 = specific enthalpy of a cooled refrigerant entering evaporator in kJ/kg.
$\dot{m}$ = refrigerant mass flow rate in kg/s.
Compressor work

$$W_c = \dot{m}(hc_2 - hc_1) \text{ in kW} \tag{2}$$

Where,
W_c = compression work input.
hc_2 = specific enthalpy of vapor exiting compressor in kJ/kg.
hc_1 = specific enthalpy of vapor entering compressor in kJ/kg.
$\dot{m}$ = refrigerant mass flow rate in kg/s.
Coefficient of performance

$$COP = \frac{\dot{m}(he_1 - he_4)}{\dot{m}(hc_2 - hc_1)} \tag{3}$$

Where,
COP = coefficient of performance.
Q_e = heat of evaporator in kJ/kg.
W_c = Compressor work done in kJ/kg.
$\dot{m}$ = mass flow rate of the refrigerant in kg/s.
Refrigerating effect

$$R.E = COP \cdot W_c \tag{4}$$

Where,
$R.E$ = refrigerating effect.

COP = coefficient of performance.
W_c = compressor work done on the working fluid.
Heat rejected by the condenser

$$hcond = \dot{m}(hc_2 - hc_3) \text{ in kW} \quad (5)$$

Where,
h_{cond} = heat of condenser.
hc_2 = specific enthalpy of vapor entering condenser in kJ/kg.
hc_3 = specific enthalpy of subcooled refrigerant exiting condenser in kJ/kg.
Refrigerant mass flow rate.

The refrigerant mass flow rate is defined as the ratio of the refrigerating effect to the enthalpy change in the evaporator.

$$Refrigerant\ Mass\ flow\ rate\ (\dot{m}) = \frac{cooling\ load}{he_1 - he_4} \text{ in kg/s} \quad (6)$$

9. Result and discussions

Figures 6, **9**, and **10** explain the thermodynamic effect that occurs in the refrigeration system. **Figure 9** shows the system's evaporating temperature (ET) when working with R600a and R134a refrigerants. The system attained its ET, the minimum operating temperature of −26°C while working with R600a in 3 hours compared to when the system worked with R134a and attained ET of −22°C in 5 hours.

Figure 6 displayed the hermetic compressor's power when the system worked with R600a and R134a refrigerant. It was clear that the system's energy consumption rate using R600a was 23.3% lower than when the system worked with R134a. The result obtained means the refrigeration system performed excellently with R600a refrigerant at ET of −26°C.

Figure 10 shows the domestic refrigeration system's coefficient of performance when working with R600a and R134a refrigerants. The result indicates that the system has a better working operation when working with R600a refrigerant. That is, the COP increases by 27.1% compared to when it worked with R134a.

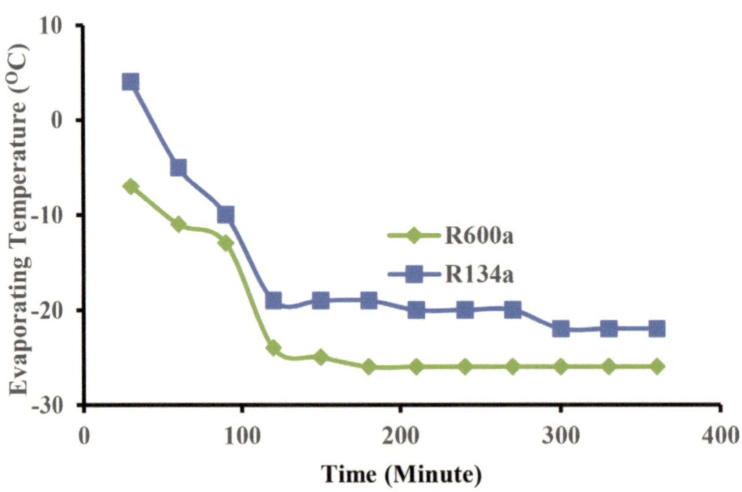

Figure 9.
Variation ET of R600a and R134a with the time taken.

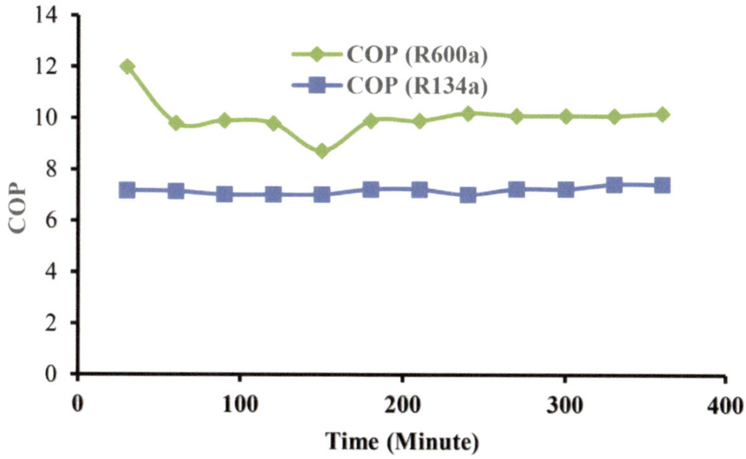

Figure 10.
Coefficient of performance with time.

10. Conclusion

This study provides access for the selection of working fluid (refrigerant), which would be appropriate for a domestic refrigerator and air-conditioning system. The effect of the working fluid contributes to the exchange rate of heat transfer in the cooling system. It also enables the computation of the system's performance characteristics (PC), which primarily predicting the system's workability and the evaporator temperature, that is, the minimum temperature required for the refrigeration system to be satisfied as efficient. The system analysis will serve as a standard measure for the procurement of any refrigeration system. Furthermore, the performance of a refrigeration system depends on the working fluid that runs through it. Therefore, based on the comparative performance of R600a and R134a refrigerants in the vapor compression refrigeration system, the working fluid dramatically impacts the overall efficiency and refrigeration effect of the cooling system, with R600a a preferable refrigerant. The use of isobutane secures the surroundings from global warming and ozone depletion and improves energy conservation.

11. Recommendation

The recommendations are based on the system prototype, a specific refrigeration system suitable for Nigeria, and other areas in the perspective of the global economy.

Prototype of the refrigeration system with relevant costing based on the economy of Nigeria. The refrigeration system made from locally sourced materials was used (see **Table 3**).

Isobutane was used as a replacement for a conventional refrigerant because of its thermodynamic properties. The refrigerant has a negligible global warming potential, zero ozone depletion potential, miscible with oil, and high critical temperature, which enhance the refrigeration system's performance [72]. The price of the hydrocarbon refrigerant (isobutane) and the refrigerator's component was suitable to employ within Nigeria. The vapor compression refrigeration system is recommended for household usage because it works using non-toxic refrigerant, as in the vapor absorption system. Likewise, thermoelectric refrigeration is less

S/N	Component	Specification	Cost (#)
1	Marine board	Plywood	15,000
2	Aluminum	Tin foil (<0.2 mm)	10,000
3	Styrofoam	EPS-thermal	3,000
4	Hermetic sealed compressor	1/12 (60–70 W)	13,000
5	Condenser	Air cooled type	1,900
6	Evaporator	1/4 Copper pipe	7,500
7	Capillary tube	Copper tube type	500
8	Refrigerant	R600a	600

Table 3.
Materials used for the construction of the refrigerator with cost in Nigeria.

efficient compared to VCRs [73, 74]. Since most developing nations have adapted isobutane, the recommendation is made for the under-developed countries to use hydrocarbon refrigerant in their refrigeration system as it has been proved eco-friendly. The use of HCs refrigerant would reduce the overall price of the refrigerator and maintenance cost in terms of power consumption because isobutane refrigerant saves energy compared to the conventional refrigerant, but this is subject to the climate conditions of various countries [75].

The overviewed recommendation for other regions in the world from the global economy perspective was that there are approximately five different climates: temperate, dry, tropical, polar, and continental. The regions with a high degree of temperature can use conventional refrigerants if they could not meet the conditions for maintaining hydrocarbon refrigerant [21]. However, countries with cool and cloudy climates could subscribe to hydrocarbon refrigerants because their temperatures are under control. Furthermore, the cost evaluation for the prototype was 15% lower than the recent cost price of a domestic refrigerator of the same volume capacity and power rating (see **Table 3**).

Acknowledgements

The authors acknowledge the Covenant University management and Centre for Research Innovation and Discoveries (CUCRID) contribution to support the implementation of this work.

Conflict of interest

The authors declare there are no conflicts of interest.

Author details

Solomon O. Banjo[1,2]*, Bukola O. Bolaji[2], Oluseyi O. Ajayi[1] and Olatunde A. Oyelaran[2]

1 Department of Mechanical Engineering, Covenant University, Ota, Ogun State, Nigeria

2 Department of Mechanical Engineering, Federal University, Oye-Ekiti, Ekiti State, Nigeria

*Address all correspondence to: solomon.banjo@covenantuniversity.edu.ng

References

[1] Banjo, S.O., Ajayi, O.O., Bolaji, B.O., Emetere, M.E., Fayomi, O.S.I., Udoye, N. E., Olatunde, A.O., Akinlabu, K. Evaluation analysis of a developed solar refrigerator using conventional refrigerant for rural and medical applications. IOP Conf. Ser.: Earth and Environmental Science. 2021; 665, 012028.

[2] Donald, B. and Nagengast. Heat and cold mastering the great indoors. ASHARE, Inc. 1791. Tullie circle NE Atlanta, GA 30329. 1994; ISBN: 1-883413-17-6.

[3] George C. and Briley P.E. A history of refrigeration. AHARE Journal. 2004; 31-34.

[4] Austin, N. and Senthil, K.N. (2012). Thermodynamic optimization of household refrigerator using propane-butane as mixed refrigerant, international journal of engineering research and applications. 2012; 2, 6: 268-271.

[5] Makhnatch, P, Mota-Babiloni, A. and Rogstam, J. Retrofit of lower GWP alternative R449A into an existing R404A indirect supermarket refrigeration system, International Journal of Refrigeration. 2017; 26: 184-192.

[6] Sadollah, E., Anwar Gavili, M.L., Taghi, D., Iraj, H. and Jamshid, S. New Class of Coolants: Nanofluids, Cutting Edge Nanotechnology, Dragica Vasileska (2nd ed.). 2010; 12-28.

[7] Desai, P.S. Modern Refrigeration and Air Conditioning for Engineer's (Principle, Practices and Application), 1st Ed. 2007; Khanna Publishers, 2-B, Nath Market, Nai Sarak, Delhi-110006.

[8] Yusuf, A.C. and Michael, A.B. Thermodynamics an Engineering Approach, 7th Edition, 20, 2011; McGraw-Hill Publishers, 1221 Avenue of Americas, New York NY100.

[9] Jingyi, M.O., Robert, D. G., McCartney, G., Enyu, G., Bent, O.J., Gerard van Dalen Schuetz, O.P., Rockett, P., Lee, P.D. Ice crystal coarsening in ice cream during cooling: A comparison of theory and experiment. 2019; 9, 6: 321, DOI:10.3390/cryst9060321

[10] Duan, Z., Zhan, C., Zhao, X., Dong, X. Experimental study of a counter-flow regenerative evaporative cooler. Building and Environment. 2016; 104: 47-58, DOI:10.2016/j.buidenv.2016.04.029.

[11] Porumb, B., Unguresan, P., Tutunaru, L.F., Serban, A., Balan, M. A review of indirect evaporative cooling technology. Energy Procedia. 2015; 85: 461-471, http://creativecommons.org/licenses/by-nc-nd/4.0/

[12] Kumar, A. and Devesh, K. Design and development of solar thermoelectric air cooling system. Imperial Journal of Interdisciplinary Research. 2016; 2, 6: 1529-1533.

[13] Shetty, N., Soni, L., Manjunath, S. and Rathi, G. Experimental analysis of solar powered thermoelectric refrigerator. International Journal of Mechanical and Production Engineering. 2016; 4, 8: 1-4.

[14] Chakraborty, J. and Bajpai, V.K. A review paper on solar energy operated vapour absorption system using Libr-H_2O. International Journal of Engineering Research and Technology. 2013; 2, 8: 1-3.

[15] Bornare, T., Badgujar, A. and Natu, P. Vortex tube refrigeration system based on compressed air. International of Mechanical Engineering and Technology. 2015; 6, 7: 97-102.

[16] Kitrattan B., Aphornratana S., Thongtip T. The performance of steam

ejector refrigeration system based on alternative analysis. Energy Procedia. 2017; 482-487, 10.2016/j. egypro2017.10.230

[17] Shuo, Z., Xianmin, G., Liang Z. Experimental study of air cycle refrigeration system for quick-freezing. International Conference on Cooling and Heating Technologies. 2012; 1-9, http://www.researchgate.net/publication/270823569

[18] Razzaq, M.A., Khan, M.M., Ahamed, J.U. Irreversibility of analysis of a split type air conditioner using R600a as refrigerant. International Conference of Mechanical Engineering and Renewable Energy. 2017; 1-6.

[19] Sawant, A.P., Agrawal, N., Nanda, P. Performance assessment of an evaporative cooling-assisted window air conditioner. International Journal of Low-Carbon Technologies. 2012; 7: 128-136.

[20] Raiyan, M.F., Ahamed, J.U., Rahman, M.M., Salam, M.A. Performance and exergetic investigation of a domestic split air conditioner using blends of R22 and R290. International Journal of Automotive and Mechanical Engineering. 2017; 14, 2: 4125-4139.

[21] Bilter. Refrigerant Report. Bilter International, Latest Edition. 2020; 71065 Sindelfingen, Germany.

[22] Homod, R.Z. Review on HVAC system modelling types and the shortcomings of their application. Journal of Energy. 2013; 1-10, DOI: 10.1155/2013/768632.

[23] Tashtoush, B., Tahat, M., Al-Hayajneh, A., Mazur, V.A. Probert, D. Thermodynamic behavior of an air conditioning system employing combined evaporative-water and air cooler. Applied Energy. 2001; 70: 305-319.

[24] Bolaji. Theoretical assessment of new low global warming potential refrigerant mixtures as eco-friendly alternatives in domestic refrigeration systems, Scientific African. 2020; 10, e00632: 1-11.

[25] Banjo, S.O., Bolaji, B.O., Ajayi, O.O., Olufemi, B.P., Osagie, I. and Onokwai, A.O. Performance enhancement using appropriate mass charge of R600a in a developed domestic refrigerator, IOP Conf. Ser.: Earth and Environmental Science 2019; 331, 1: 1-8.

[26] Basha, T.M., Ranganna, H., Yadav, G. M.P. Optimum length of a condenser for domestic vapor compression refrigeration system, International Journal of Science, Engineering and Technology Research. 2015; 4, 2: 277-281.

[27] George, C. and Briley, P.E. A history of refrigeration. AHARE Journal. 2004; 31-34.

[28] Adekunle, A., Arowolo, T.A., Adeyemi, O.M., Kolawole, O.A. Estimation of thermal comfort parameters of building occupants based on comfort index, predicted mean vote and predicted percent of dissatisfied people in the northern west zone of Nigeria. International Journal of Advanced in Engineering and Management. 2020; 2, 5: 809-826.

[29] Orhewere, B.A., Oluseyi, O. A., Solomon, O.B. and Ajayi A.A. Data on the no-load performance analysis of a tomato postharvest storage system (data in brief), IOP Conference Series Material Science and Engineering. 2017; 13: 667-674.

[30] Bhavesh, A.K., Mohod, S.A. and Dhawade, M. Effect of different refrigerants on vapor compression refrigeration system—A review. International Journal of Engineering Development and Research. 2016; 4, 2: 329-335.

[31] McMullan, J.T. Refrigeration and environmental issues and strategies for

the future. International Journal of Refrigeration. 2002; 25: 89-99.

[32] Ibhabode, O., Adehunle, A.A., Banjo, S.O., Atakpu, O.D. Thermophysical, electrical and mechanical characterizations of normal and special concrete: A holistic-empirical investigation of pre-qualification and quality-control of concrete. Journal of Physics. 2019; 1378: 1-76, doi:10.1088/1742-6596/1378/4/042100.

[33] Ishak, F., Dincer, I., Zamfirescu, C. Thermodynamics analysis of ammonia-fed solid-oxide fuel cell. Journal of Power Sources. 2012; 202: 157-165.

[34] Bolaji, B.O. Experimental study of R152a and R32 to replace R134a in a domestic refrigerator. Energy. 2010; 35: 3793-3798.

[35] Akash, B.A. and Said, S.A. Assessment of LPG as a possible alternative to R-12 in domestic refrigerators. Energy Conversion and Management. 2003; 44:381-388.

[36] Vaibhav, J. Kachwaha, S.S. and Mishra, R.S. Comparative performance study of vapour compression refrigeration system with R22/R134a/R410A/R407C/M20. International Journal of Energy and Environment. 2011; 2, 2: 297-310.

[37] Ranendra, R. and Bijan, K.M. First law and second law analysis of mechanical vapour compression refrigeration system using refrigerants CFC12, R134a, R290. International Journal of Current Engineering and Technology. 2014; 3: 191-195.

[38] Fernandez-Seara, J., Uhia, F.J., Diz, R. and Dopano, J.A. Vapour condensation of R22 retrofit substitutes R417A, R422A and R422D on Cu.Ni turbo C tubes. International Journal of Refrigeration. 2010; 33: 148-157.

[39] Eastop and McConkey, Applied Thermodynamics for Engineering Technologist. 2009; New Delhi 110 017, Indian.

[40] Banjo, S.O., Bolaji, B.O., Osagie, I., Fayomi, O.S.I., Fakekinde, O.B., Olayiwola, P.S., Oyedepo, S.O. and Udoye, N.E. Experimental analysis of the performance characteristics of an eco-friendly HC600a as a retrofitting refrigerant in a thermal system, Journal of Physics. 2019; 1378, 4: 1-8.

[41] Bolaji, B.O. Selection of environment-friendly refrigerants and the current alternatives in vapour compression refrigeration system. Journal of Science and Management. 2011; 1, 1: 22-26.

[42] Dalkilic, A.S. and Wongwises S. A performance comparison of vapour compression refrigeration system using various alternative refrigerants. International Communication in Heat and Mass Transfer. 2010; 7: 1340-1349.

[43] Ajayi, O.O., Useh, O.O., Banjo, S.O., Owoeye, F.T., Attabo, A., Ogbonnaya M., Okokpujie, I.P. and Salawu, E.Y. Investigation of the heat transfer effect of Ni/R134a nanorefrigerant in a mobile hybrid powered vapour compression refrigerator. IOP conference series. Materials Science and Engineering. 2018;391: 1-8.

[44] Reddy, D.V.R., Bhramara, P., Govindarajulu, K. Performance analysis of domestic refrigerator using hydrocarbon refrigerant mixtures with ANN and fuzzy logic system. International Conference Numerical Heat Transfer and Fluid Flow. NIT; 19-21 January 2018; Warangal, India: p. 1-7

[45] UNEP, United Nation Environment Program, Handbook for International Treatics for Protection of Theozone Layers, 5th ed. 2003; Narobi, Kenya.

[46] Adekunle, A., Osagie, I., Ibhadode, A.P., Caesar, S.M. Assessment of carbon emissions for the construction of buildings using life cycle analysis: Case study of Lagos state. International Journal of Engineering Research and Advance Technology. 2020; 6, 8: 1-11.

[47] Prakash, U, Vijavan, R. and Vijay, P. Energy and exergy analysis of vapour compression system with various mixtures of HFC/HC. International Journal of Engineering, Management and Applied Sciences. 2016; 4, 1: 40-48.

[48] Tashtoush, B. Tahat, M. and Shudeifat, M.A. Experimental study of new refrigerant mixtures to replace R-12 in domestic refrigerator. Applied Thermal Engineering. 2002; 22: 495-506.

[49] Somchai, W. and Nares, C. Experimental study of hydrocarbon mixtures to replace HFC-134a in a domestic refrigerator. Energy Conversion and Management. 2005; 46: 85-100.

[50] Sekhar, S.J. and Lal, D.M. HFC134a/HC600a/HC290 mixture a result for CFC-12 system, International Journal of Refrigeration. 2005; 28: 725-743.

[51] Dreyfus, G., Borgford-Parnell, N., Fahey, D.W., Motherway, B., Peter, T., Picolotti, R., Shah, N., Xu Y. Assessment of climate and development benefits of efficient and climate-friendly cooling. 2020; 1-89.

[52] Mohanraj, M., Jayaraj, S. and Muraleedharan, C. Environment friendly alternatives to halogenated refrigerants—A review. International Journal of Greenhouse Gas Control. 2009; 3: 108-119.

[53] Peyyala, A and Sudheer, NVVS. Experimental investigation of COP using hydrocarbon refrigerant in a domestic refrigerator. IOP Conference Series. Materials Science and Engineering. 2017; 225: 012236.

[54] Sanket, B. Substitution of non-ecofriendly refrigerants by hydrocarbon refrigerants: A review. International Journal of Advance in Science Engineering and Technology Aurangabad. 2016; 2: 122-125.

[55] Kundu A., Kumar, R. and Gupta, A. Performance comparison of zeotropic and azeotropic refrigerants in evaporation through inclined tubes. 1ith international conference on mechanical engineering, ICME, Procedia Engineering. 2014; 90: 452-458.

[56] Qureshi, M.A. and Bhatt, S. Comparative Analysis of COP using R134a & R600a refrigerant in domestic refrigerator at steady state condition. International Journal of Science and Research (IJSR), Prestige Institute of Engineering and Science. 2014; 3, 12: 935-939.

[57] Ahamed, J. U, Saidur, R, Masjuki, H. H. and Sattar, M.A. An analysis of energy, exergy and sustainable development of a vapor compression refrigeration system using hydrocarbon. International Journal of Green Energy, 2012; 9, 7: 702-717.

[58] Banjo, S.O., Fayomi, O.S.I., Atayero, A.A.A., Bolaji, B.O., Dirisu, J.O., Okeniyi, J., Emetere, M.E., Olorunfemi, B.J., Owoeye, T. Effect of fins spacing on the performance evaluation of a refrigeration system using LPG as refrigerant. IOP Conf. Ser.: Earth and Environmental Science. 2021; 665, 012030.

[59] Harby, K. Hydrocarbon and their mixtures as alternative to environmental un-friendly halogenated refrigerants: An updated review. Renewable and Sustainable Energy Reviews. 2017; 73: 1247-1264.

[60] Sattar, M.A., Saidur, R. and Masjuki, H.H. Performance investigation of domestic refrigerator using pure hydrocarbons and blends

hydrocarbons as refrigerants. World Academic of Science Engineering and Technology. 2007; 5: 223-228.

[61] Bolaji, B.O. and Huan, Z. Ozone depletion and global warming: Case for the use of natural refrigerant—A review. Renewable and Sustainable Energy Reviews. 2013; 18: 49-54.

[62] ASHARE, Air-Conditioning and Industry Refrigerant: Refrigeration Selection Guide. The Australian Institute of Refrigeration, Air-Conditioning and Heating (AIRAH). 2003; Australia.

[63] Dhamneya, A.K., Rajput, S.P.S and Singh, A. Comparative performance analysis of ice plant test rig with TiO_2-R-134a nano refrigerant and evaporative cooled condenser. Case Studies in Thermal Engineering. 2018; 11: 55-61.

[64] ASHRAE, Designation and Safety Classification of Refrigerant. American Society of Heating, Refrigerating and Air-conditioning Engineers, Atlanta, GA, USA; 2001.34.

[65] Bolaji, O.O. Energy performance of ecofriendly R-152a and R600a refrigerants as alternative to R134a in vapour compression refrigeration system. Analele UniversitaTii "Eftimi Murgue" Resita. 2014; 1: 354-367.

[66] Garland N.P. and Hadfield M. (2005). Environmental implication of hydrocarbon refrigerants applied to the hermetic compressor. Materials and Design, 26, 578-586.

[67] Raut, P.D. Study of refrigeration system for achieving cryogenic temperature. International Journal of Science and Research, 2017; 6, 5: 1444-1448.

[68] Mokhatab, S. and Poe, W.A. Handbook on Natural Gas Transmission and Processing. Golf Professional Publishing. 2012; Burlington, MA, USA.

[69] Senthlkumar, D. Influence of cryogenic treatment on TiC namopower in R600a and R290 refrigerant used in vapor compression refrigeration system. International Journal of Air Conditioning and Refrigeration. 2019; 27, 4: 155040

[70] Kalantar-Neyestanaki, H., Mafi, M., Ashrafizadeh, A. A novel approach for operational optimization of multi-stage refrigeration cycles in gas refineries. International Journal of Refrigeration. 2017; 80: 169-181.

[71] UNEP/ASHRAE. Update on new refrigerants designations and safety classification. 2020; UNEP/ ASHRAErefrigerantfactsheet.https:// www.ashrae.org/file%20library/technical%20resources/refrigeration/factsheet ashrae_english-20200424.pdf.

[72] Abas, N., Kalair, A.R., Khan, N., Haider, A., Saleem, Z. (2018). Natural and synthetic refrigerant, global warming: A review. Renewable and Sustainable Energy Reviews, 90, 557-569.

[73] Shetty, N., Soni L., Manjunath S., Rathi G. Experimental analysis of solar powered thermoelectric refrigerator. 2016; 4, 8: 99-102.

[74] Pearson, A. Ammonia as a refrigerant. International Institute of Refrigeration, 2008; Paris France.

[75] Ciconkov, R. Refrigerants: There is still no vision for sustainable solution. International Journal of Refrigeration, 2018; 86: 441-448.